THE ROBOTS ARE COMING

THE ROBOTS ARE COMING

A Human's Survival Guide to Profiting
in the Age of Automation

John Pugliano

Ulysses Press

Published in the United States by:
Ulysses Press
P.O. Box 3440
Berkeley, CA 94703
www.ulyssespress.com

ISBN: 978-1-61243-669-2
Library of Congress Control Number 2016957546

Printed in Canada by Marquis Book Printing
10 9 8 7 6 5 4 3 2 1

Acquisitions: Casie Vogel
Managing editor: Claire Chun
Editor: Renee Rutledge
Proofreader: Shayna Keyles
Indexer: Sayre Van Young
Front cover and interior design: what!design @ whatweb.com
Cover photos: robot © Ociacia/shutterstock.com; man © SFIO CRACHO/
    shutterstock.com
Layout: Jake Flaherty

Distributed by Publishers Group West

This book is dedicated to my paternal grandparents, Antonio and Maria, who were born in a remote Italian village in the early 1880s. They were able to rise up from near-medieval poverty because of the amazing technological innovations of the twentieth century. May we all live such long and fascinating lives.

CONTENTS

INTRODUCTION

If you are like most readers, you might gloss over a book's introduction or skip it altogether. This book is unique, and I advise you to read the introduction. *The Robots Are Coming* is not a prediction of future events, nor is it simply an assessment of automation's impact on our lives. The book is written to serve as a guide or manual. It is an interactive document with actionable instruction designed to help you survive and navigate through the tumultuous robotic times ahead. Uniquely, the topics discussed will be from the perspective of employment, as well as taking into account economic and investment concerns.

The best way to use this book requires your active participation. Throughout the book, I have provided Action Plans that will help you build your own survival plan. The process laid out in this book is meant to be iterative rather than a quick and simple informational read. Take the time to read the content, ponder its implications on your personal situation, and then take action to prepare yourself for the impact of an automated world. Reread and reconsider as necessary.

Professionals, Take Heed

Some inaccurately believe that automation will disproportionately have a negative impact the working class blue collar employee. The premise of this book is that over the past generation, those labor-intensive blue collar jobs have already been discounted by automation. The real bite of the next round of automation will be felt by the previously insulated white collar workers, like middle management, legal, and medical professionals.

Higher income earners that have so benefited from the efficiencies of the information age will soon find their services in direct competition with the next wave of technology. Big data, advanced algorithms, inexpensive sensors, and robotics will all converge to tackle the lucrative jobs of the white collar professional. Any job function that is routine and predictable will be a target for efficiency improvements through automation. Automation will aggressively supersede the work of highly compensated professionals because replacing human labor in those jobs will provide the highest return on investment. Yes, labor-saving devices will replace employees at fast food restaurants, and society's budget-cutters will invest in technology that makes medical professionals who earn $300,000 salaries redundant.

No Crystal Ball

We cannot predict the future. The best we can hope for is to anticipate and then adapt and overcome. Throughout this book, I have used historical references as a basis for assessing future outcomes. Exactly what technologies will be developed and how quickly society will adopt them is uncertain. Logic would indicate that both development and adoption rates will continue to increase, as they have since the Industrial Revolution. So the impact will likely be sooner rather than later.

Some assumptions about the future must be considered as the basis for formulating a starting point. However, the intent of this book is not to predict which technologies will prevail. The value of the book's insight is to help you develop survival strategies for the inevitable economic changes brought on by automation, regardless of the specific technology employed.

While future technologies will be discussed in this book, its emphasis is on mankind, not the machine. For while we cannot predict the future, we can with some certainty predict people's actions. Human characteristics, such as love, hate, fear, and greed, appear to be uninfluenced by technological change. As such, we will explore what I have found to be the historical constant and future solution: your unique humanity, or human touch.

How to Use This Book

This book is comprised of four parts, each beginning with a chapter that challenges you with a cognitive instruction:

- Think like a human, not a machine.

- Think like an entrepreneur, not an employee.

- Think like a saver, not a consumer.

- Think like an investor, not a speculator.

Each chapter ends with an Action Plan to help you consider how automation might be a threat and to provide coping responses. Use these exercises to help you think in economic rather than emotional terms. The economic reality of the coming automation revolution is that the robots are coming to take your job. Ultimately, this fact will not be altered by emotional response, political policy, or unionized negotiation. If you want to remain competitive in the face of automation, it cannot be done simply through productivity. A human cannot outperform a robot at a repetitive task. The robot will eventually win.

Your competitive edge must come from filling an economic niche that is based on your human touch. Use the Action Plans as your template for aligning realistic market needs with your unique talents. Out of necessity, the exercises are generic in nature, with open-ended questions that can apply to a broad audience, equally applicable to the carpenter or the cardiologist.

The reader is ultimately responsible for drilling down to specificity, because that is the way the real world works. There are no cookie-cutter answers. The harder that you work to answer and adapt the Action Plan questions to your own situation, the more likely you will find a viable solution to your unique place in the robotic future. Ultimately, your place in the future can only be determined by you. I cannot know what is right for you, nor can anyone else. It is a personal journey that is your responsibility. As an author, I can act only as a guide and encourage you to think for yourself; thus, the four cognitive instructions that begin each section.

Corps of Discovery

The Corps of Discovery was the official name of the Lewis and Clark Expedition of 1804, which you are probably very familiar with. Lewis and Clark were commissioned by President Thomas Jefferson to explore the newly acquired territory of the Louisiana Purchase. A key objective was to locate a navigable water route across the continent to the Pacific Ocean.

Believe it or not, the Corps' expedition is specifically relevant to your journey into the unknown robotic future for three reasons.

First, Lewis and Clark had no specific map to follow. The belief of the time was that the Missouri River bridged the continental gap to the Pacific coast, but no one knew for sure. Since the Corps had no specific map to follow, they could only pursue a general course and hope for the best. To improve their odds of success, they prepared for the trip by developing useful skills: navigation, bush craft, medicine, and scientific discovery methods. Likewise, you should prepare for your uncharted course into the future by setting out in a general direction accompanied by useful core skills.

Second, the intention of the expedition was for commercial purposes. Today, we romantically remember Lewis and Clark as glamorous explorers and credit their many scientific discoveries, but Jefferson specifically commissioned the Corps to find ways to commercially exploit the region.

Your journey should similarly include the long-term goal of building wealth, thus this book's emphasis on economic and investment matters.

Third, the expedition was graced with good luck. In addition to the preparation and skill of the men, good fortune played a decisive role. Stranger than fiction, Sacagawea's involvement is one of a fairy tale heroine. Sacagawea joined the Corps as the pregnant companion of a French trapper. She had been kidnapped as a child, sold into slavery, and eventually became the teenage "wife" of Charbonneau, the French trapper. Lewis and Clark considered her ability as an interpreter and many other skills as essential to the Corps.

Your success will be largely determined by your level of preparation and personal skills. Regardless of your preparation and efforts, luck often plays a big role in any journey. Navigating the robotic future will be no exception. However, if you're prepared to leverage your skills and understand what is at stake, you'll be able to recognize your own Sacagawea when she appears in your story.

Like Lewis and Clark, you are embarking on your own journey of discovery. Use this book as a guide to help you anticipate future trends and to adopt innovative technologies that complement your talents. Make the effort to complete the Action Plans and then use them as a template to strategically plot out your own course of action. Do not be afraid to think differently from the crowd. In fact, that is when you will know you are headed in the right direction. Nonconformity will lead you to think like a human, an entrepreneur, a saver, and an investor. Your future will be framed by your thoughts.

PART ONE
Humanity

Chapter 1

THINK LIKE A HUMAN

You might be familiar with this apocryphal quote attributed to Henry Ford: "If I had asked people what they wanted, they would have said faster horses." The essence of that statement is that progress does not originate with the consumer, but rather with the innovator. A similar line of reasoning can be applied to combating the inevitable loss of jobs to robotic automation.

If you were to ask an employer what they wanted in an employee, they would say something to the effect of "faster, cheaper, more productive." However, these are not skill sets readily attributed to humans. People get tired, bored, forgetful, emotional, and, oftentimes, they exhibit self-destructive or antisocial behavior. Machines come with none of these flaws; they just execute commands. If you were an employer, who would you hire?

You Can't Beat a Robot at Repetition

Prior to the 1990s, automobile manufacturers employed thousands of skilled workers as painters and welders. Today those assembly-line tasks are almost exclusively done with robotics. Human skill could not match the precision of an industrial robot. The change did not happen overnight: There was opposition from labor unions and the prospect of globalization, with companies offering lower wages in foreign countries. So, initially, workers went out on strike and jobs migrated overseas. Ultimately, robots prevailed because of basic economics; their skills increased, and their price decreased. And it's important to remember that a robot does not have to be a physical machine, but anything that automates a task.

There are several lessons to be gleaned from the adoption of industrial automation. First, a human cannot be more productive than a robot at a repetitive task. Period. It makes no difference what the task is. Obviously, simple tasks are easier to automate than complex ones. But as the cost of computing power decreases, more complex tasks can be reduced to a mathematical algorithm.

Consider the ancient game of chess. Arguably, there are over 1,050 possible moves in a game. Skilled human players are rare. Globally, there are only about 1,500 living players that have earned the title "grandmaster."

In 1997, the first computer was able to defeat a sitting world champion in an official tournament. The program was run on IBM's supercomputer Deep Blue. By 2005, comparable computing power was available on desktop computers. Today, similar programs can be run on a smartphone.

The Threat to White-Collar Jobs

The second lesson that can be learned from industrial automation is not as obvious as the first: Automation does not replace the simplest task, but instead, it is a compromise between complexity and cost. Think back to the automotive assembly line; the painters and welders were *skilled* workers. It would have been far easier and cheaper to build a robot to screw lug nuts on a wheel than to develop systems to replace a skilled painter or welder. Yet it was precisely the skilled labor that was targeted because of cost savings. Productivity is improved more by replacing a highly compensated skilled employee than an average line worker.

Total operating costs account for products that must be rejected or reworked due to poor quality. Botched paint jobs or poorly welded joints are more costly to correct than a cross-threaded bolt. So replacing a skilled worker with a precision robot is a win-win for the employer: it lowers operating costs and improves quality.

Skilled craftsmen and white-collar professionals are not immune from receiving a pink slip. Quite the opposite, high-income earners that perform repetitive tasks are the most likely victims of automation.

EXTINCTION OF THE MIDDLE MANAGER

If I had to pick one career that would be most impacted by automation, it would be the proverbial "middle manager." It might not have been obvious, but they have been in a death spiral for decades. Up until this point, their demise has moved at the speed of a glacier. Their final chapter will close swiftly and definitively, like an avalanche. The reason is obvious: a middle manager's job function is complex, yet extremely routine. Enterprise software has been nibbling away at the mid-level manager's role since the latter part of the twentieth century. Think of the success of companies like Oracle, SAP, and Salesforce. The middle managers at Fortune 500 companies that have so diligently implemented these programs have been training their replacements. A huge profit windfall will occur by eliminating the middle hierarchy of white-collar managerial jobs once enough historic data has been collected and correlated to be fully operational by enterprise software.

So, Are We All Doomed?

If a corporate middle manager with an MBA education is not secure in employment, what chance do you have? More than you think, as long as you start thinking like a human and not like a machine. A robot is the proverbial "cog in the wheel." It performs the task it has been pro-grammed to do. Nothing more. As a human, you have unique insight and creativity that cannot be programmed, because it does not exist until you create it. The key is the human element of creativity, but more about that in later chapters.

WHAT DO AMATEUR (HAM) RADIO OPERATORS AND PORTUGUESE WATER DOGS HAVE IN COMMON?

There are more of them now than at any time in history.

It defies logic that people would want to use archaic com-munication technology or that urban dwellers would own big, hairy fishing dogs. The point is that humans are not logical, nor do they act in ways that can be predicted by linear models. Ubiquitous smartphones and voice over Internet protocol (VOIP) did not kill modes like Morse code; quite the opposite. Internet, digital, and satellite communi-cation systems have enhanced the ham radio experience. The hobby has never been more popular.

The same concept applies to Portuguese water dogs and other working breeds, like shepherds. These dogs have been bred for centuries for the specific purpose of assist-ing fishermen along the coast of Portugal. Today, they are prized for their loyalty, companionship, and hypoallergenic fur. President Obama owned two of them while in the White House. He did not need a dog breed for fishing, but his children wanted one (or two).

Americans spend nearly $60 billion dollars per year on their pets. For the most part, ownership is purely nonutil-itarian. Dogs are not used for security or herding; cats are not used for vermin control. People simply love their pets and regard them as members of the family. This is not rational, in an economic sense. However, it is a predicable human characteristic that resonates from our primordial history. It is a characteristic than can never be captured by an algorithm or mimicked by a robot.

So, do not despair, there is a bright future for those who improve and monetize their unique human traits.

Below is a list of the most relevant attributes needed for the future economy.

Traits for the Future Economy

I focus here on traits or attributes rather than specific job functions or skills for several reasons. A worker in the 1950s employed in the printing industry would have limited job opportunities today because that industry has mostly been made obsolete by digital technology. However, that worker may have developed crossover traits that still have value in the future economy. Similarly, the typing speed of a 1960s-era office worker is not very relevant today, but her traits of organization and communication would be.

The following traits are broadly descriptive. For example, "electrical" describes a career trait that will be highly useful in the future economy, yet it is broad enough to encompass a wide range of careers: electrical engineer, electronic technician, electrician, self-taught maker, and even job titles that do not currently exist.

The leading traits here are followed by supporting traits. This is important because while all of the traits have value, some cannot stand alone. In fact, all the traits will be most effective when combined with others. For example, Elon Musk is simultaneously the founder of Tesla, SolarCity, and SpaceX. An entrepreneur of his stature obviously is not vocationally one-dimensional. He most likely has these traits: digital thinker, mechanical attributes, electrical knowledge, competent, organized, courageous, etc. Musk is a master of many traits, which is why he is a billionaire. You may initially only identify one leading and one supporting trait that relate to you. That is okay! The important thing is to get started.

Leading Traits

Digital Thinking

Digital thinking is a trait beyond the traditional field of computer science. Yes, it involves programming, software, and hardware, but it also encompasses a different way of thinking. Digital thinking is the realm of ones and zeros, on or off. Rule-based logic is at the basis of digital thinking: If a=b and b=c, then a=c. It permits finite reality to be expressed and emulated in near-infinite terms. From artificial intelligence to virtual reality, digital thinking will be the cornerstone of all future actions. Digital thinking is the catalyst that will transform today's limited analog world into tomorrow's perpetual abundance.

Mechanical Attributes

Mechanical attributes will be crucial if for no other reason than the fact that a proliferation of robots will mean lots of moving parts. The robotic lifecycle will require human tending: design, installation, maintenance, reprogramming, decommissioning, and disposal. Robots will do much of the work, but they will never be completely independent of human oversight.

Electrical Knowledge

Electrical knowledge will go hand in hand with mechanical attributes. Electronics have a higher reliability than mechanized functions, so maintenance requirements will not be as great. The majority of electrical work will take place during the initial stages of manufacturing, assembly, and installation. The good news for people that possess electrical knowledge is that electronic components will literally be omnipresent. Almost every conceivable device will be coupled to the Internet. Their failure rate will be low but they will be everywhere, so from a sheer volume standpoint, these devices will be a large percentage of the economy.

Biological Expertise

Advances in genetics and biochemistry have revolutionized the field of traditional biology. For those interested in the life sciences, opportunities will abound. However, this attribute, more than the other leading traits, will require an interdisciplinary approach. To unlock the potential within the biological field, expertise in digital, mechanical, or electrical fields will be a winning combination. Think technologies like genetic engineering, prosthetics, implants, bionics, cyborgs, and space travel. The possibilities are truly endless.

Supporting Traits

Kind

Kindness may seem inconsequential in an automated world; however, I would argue it will be a trait in high demand. In a world where people have the choice to select human interaction over automated processes, human interaction will be more costly and, so, at a minimum, it must be enjoyable. Suppose you need to renew your state driver's license. Which would you choose? Dealing with a rude Department of Motor Vehicles worker, or online registration? Conversely, if you were the victim of a crime, would you rather fill out an online incident report or have your statement taken by a caring police officer? The economy of the future will provide consumers an abundance of low-cost products, but I think human kindness will remain a scarce resource.

Competent

In a world of robotic precision, incompetence will not be tolerated. Whatever one's job function is, the vital standard will be properly accomplishing the task. There will be plenty of work for *competent* doctors, teachers, and plumbers. Datamining, cloud computing, and social media feedback will instantaneously critique everyone's performance. Those that consistently perform poorly will find themselves unemployed or without customers. This is already occurring. During twenty years of corporate travel, I had experienced extremely poor, overpriced service from taxicabs all around the world. With the exception of that in Hong

Kong, I would categorically describe the taxi industry as consisting of old, dirty vehicles and incompetent drivers. In recent years, I have had exactly the opposite experience with Uber. Before I ever enter an Uber car, with the use of a simple free smartphone app, I can predict the entire driving experience. I know who the driver is, his rating, the type of vehicle, the fee, pickup time, route, and time to destination. Uber is competent and, when given a choice, that is the service that I hire.

Communication Skills

The ability to communicate has always been and will remain an essential function for success. Think of communication in a broad sense as accurately conveying information in two directions. It consists of transmission to a target audience and reception back, for completion of a full communication cycle. Communication is complex, extending beyond human language and culture to dialog with computers and robots. A best overall practice for success is to first select your primary area of interest (say, electrical knowledge or digital thinker) and then develop a communication expertise customized to the primary audience. The communication plan might require learning a programming language, a foreign language, or, perhaps, learning to communicate with small children. It all depends on the needs of your primary target audience. The future economy will need smooth-talking marketing types, as well as techno geeks that speak machine language.

Artistic

The importance of art may seem somewhat illogical in the age of automation because art is pure emotion, speaking to the human and not the machine. In fact, art will remain relevant for the very reason that it will provide an emotional counterweight to the sterile world of technology. Art is closely related to, if not a direct subset of, communication. So, think of artistic traits in terms of communicating on an emotional level by using the senses of sight, touch, hearing, smell, and taste. It is difficult to succeed solely on the basis of artistic ability (you have heard of the "starving artist"), so strive to incorporate art into other traits like digital thinker or electrical knowledge.

Organized

Organization is required to build any enterprise, from establishing a career to launching a multinational brand. Organization is required of people at all levels, from a corporate CEO to a restaurant short-order cook. Organization is manifest when logical decisions are based on reliable facts. It requires discipline and non-emotional objectivism; thus, it is very difficult for most people to execute. The organizational trait is especially powerful when it can be successfully combined with a counter trait like artistic or visionary.

Visionary

Visionary is to some degree the opposite of organized. A visionary does not ignore the established facts, but he does optimistically look beyond today's limited horizon at what might be possible in the near future. Incremental success can occur without visionary thinking, but explosive exponential growth can only occur with it. The uncertain times ahead will present the visionary with unlimited opportunity.

Courageous

The ability to determine risk and accept its consequences is among the rarest of traits. While a high level of risk tolerance has always been beneficial, I believe it will become even more important during the turbulent economic times ahead. Those who are courageous will be more likely to quickly adopt technology and react to changing conditions. It is no coincidence that entrepreneurs have a high degree of risk tolerance. A note of warning: a supporting trait should never be used as a leading trait; this is of particular importance when dealing with the courageous trait. Actions that are primarily based on a high risk tolerance quickly digress from the rational to the speculative.

Harmonize Traits

Think of the traits as musical notes. Mix and match them to fit your particular style. The sound and tempo that appeal to you might be noise to someone else. What is important is that it is true to you.

You will also find that proficiency in one trait will likely spur interest in another, possibly leading you to a previously unknown talent or interest. This is similar to the way a music student might first learn to play the piano before moving on to other instruments and finally discovering that their passion is the guitar. Progressively building on multiple traits is how one becomes a Renaissance man like Elon Musk, Thomas Edison, or Benjamin Franklin.

As a general rule, the more traits that you can incorporate into your repertoire, the more career options you will have. But you do not want to pursue a trait willy-nilly if it does not ring true to your innate abilities. Mechanical attributes will not be of benefit to you if you do not know which end of a screwdriver to use. For example, I have absolutely no sense of direction. It would be foolish for me to pursue a career as a navigator simply because it was a highly compensated job. Even if I were able to find employment as a navigator, I would never be better than marginally competent. Imagine the frustration of going to work every day knowing that your best effort resulted in mediocre results. I am sure you have met many people like this that pursued a career for the income or because their parents thought it was a good idea. Those are the saddest of people.

At a minimum, try to identify at least one leading trait. The four categories should be broad enough that most people can find at least a glimmer of interest. The importance of those specific four traits is that they will play a crucial role in the robotic economy of the future. Of course, there will be other routes, but those four will offer the path of least resistance.

Likewise, it would not be impossible to build your career on one of the supporting traits, it just will not be as easy. For example, many people try to eke out a living as an artist (painters, musicians, sculptors, stand-up comedians...the list is endless). But only a minute fraction of self-proclaimed artists actually earn a full living directly from their art. Most derive the majority of their income from another source, such as teaching, bartending, etc. Their lack of success is not due to a talent deficiency. It is most likely related to the fact that the artistic trait is supporting in nature and just does not offer enough earning potential. The same is true of the other supporting traits. It is hard to earn a living if your primary function is simply being kind or courageous. Kindness is a noble virtue, it just is not a standalone income-producing attribute.

On the other hand, a person that is able to combine multiple leading and supporting traits will greatly improve the odds of their success. A high-tech corporate CEO is likely to possess at least one leading trait and several supporting traits, such as digital thinker, organized, visionary, communication skills, competent, and courageous.

A successful nurse practitioner might possess the following traits: biological expertise, organized, competent, and kind.

Trait-Based Thinking

Automation will always outperform a human at logical repetitive tasks. The good news for the human race is that we live in an illogical, unpredictable world. Those who focus on developing their unique personal human skills will prevail, while the folks trying to compete with robots will be made redundant. The key concept is to *think like a human, not a machine*.

This concept might be alien to you because we are taught from an early age to be logical and conform. Institutions (corporations, the military, and universities) like conformity. Business schools and engineering departments promote standardization. Standardization is not in and of itself more effective than randomness; however, standards can be measured. So institutions follow the path of least resistance by implementing standards to predict future outcomes.

One of the reasons I strongly predict the demise of the middle manager is that much of their responsibility is to manage the status quo, which can be reduced to an algorithm and executed by a computer decision support system. Search for roles of a manager and you will find terms like supervise, implement, communicate, train, and negotiate. A high school coder could easily write an app to carry out these functions. Absent are entrepreneurial characteristics like vision, risk aversion, and creativity. When was the last time you heard of a mid-level manager position being filled by a loose cannon with a job description of "shake things up"? Never.

Albert Einstein said, "We cannot solve our problems with the same thinking we used when we created them." Likewise, we cannot prevail over automation by thinking like a robot; we will achieve by thinking like the beings we are, creative humans.

Unconventional Innovators

Breakthrough achievement does not occur from incremental improvement on the margins. Marginal improvement, functions the robots will excel at (faster, lighter, smaller), will be the domain of automation. For humans to win, they will have to compete on a different plane. Fifty years ago, architectural firms were staffed with draftsmen to translate the ideas of architects to paper. Draftsmen have long since been replaced by CAD/CAM software. Architects still exist because their function is to create.

Creativity is not rational. It does not flow from reason or logic but from emotion. Creativity is the confluence of the rational left brain and the emotional right brain. Look to history for examples of breakthrough events that ushered in a new technological age. The innovators were mostly right-brained visionaries, not left-brained theoreticians.

Ulysses S. Grant

The US Civil War marked the transition of warfare from the provincial cavalry charge to an industrial age of mechanized weaponry. By 1863, President Lincoln was frustrated with his general staff and their inability to battle the determined South. The best generals of the North could not win against the tactics of the South under the leadership of General Robert E. Lee. Lee was a top graduate of West Point. So, after all conventional attempts failed, who did Lincoln appoint to defeat Lee? Ulysses S. Grant, who graduated in the bottom half of his class at West Point. Grant was such a poor administrator that he had been washed out of the officer's corps after a decade of service. Yet it was Grant who turned the tide against the South. Victory came 14 months after Grant was appointed General-in-Chief of the Armies.

Grant's battle victories occurred because unlike his peers, Grant ignored the fighting tactics taught at West Point. Conventional war strategy of the time taught brute force regimented formation fighting like that used by Napoleon in Europe. Grant focused on winning small skirmishes with speed and agility. To accomplish this, he readily adopted the new emerging technologies of the day—the telegraph and railroads. Grant was an innovative leader.

Henry Ford

In the early twentieth century, the automobile was a novelty. Up until the car was introduced, the steam locomotive was the most innovative mode of transportation. It took the vision of quirky Henry Ford to birth a new industry. (Historic note: Ford was born about the time Grant was defeating Lee.) Ford was a self-taught engineer with a passion for tinkering. He was a low-level employee until his mid-thirties, and was 45 when the Model T was launched. Ford was a visionary but not necessarily what could be classified as a holistic thinker. Like many inventors, he often overlooked the details of the present because his focus was on the future of his creation. In 1896, when he built his first gas-powered vehicle, it was too large to fit through the garage door.

Steve Wozniak and Steve Jobs

In the latter half of the twentieth century, a consumer-grade computer was a novelty. Business and research computing was dominated by large corporations like IBM. It took the visionary duo of college dropouts Steve Wozniak and Steve Jobs to create Apple Computer. (Historic note: Wozniak was born three years after the death of Ford.) Wozniak was the self-taught electronics engineer and Jobs was the forward-thinking genius. Jobs was such a controversial character that in 1985, he was fired from the company he founded.

Technically, Jobs was not fired; he quit after the board of directors stripped him of decision-making authority. Rather than remain as a figurehead, he resigned in protest. Jobs's confrontation with the board of directors was about long-term vision versus short-term profits. Jobs wanted to spend money developing and marketing the Macintosh concept, while the board wanted to promote the profitable Apple II. The Apple II was a powerful niche computing tool with market acceptance, while the Macintosh was seen as a "toy." Jobs did not seek compromise. He spent large sums developing the Macintosh and staffed the marketing team with "unconventional" talent. Rather than relying solely on technical staff, the marketing team included an eclectic group of artists, musicians, and poets.

A decade after his departure, the floundering, shortsighted Apple board of directors rehired Jobs to lead a turnaround. Jobs went on to not only reinvent Apple, but entire industries—computer, software, telephone, consumer electronics, and music, to name a few. Jobs was an innovative CEO.

The above examples are not unique. Unconventional characters always rise to the top during times of epoch transformation. For you to survive and thrive during the robotic revolution, you, too, must be unconventional. Robots are the cog in the wheel. To maintain your relevancy in the age of automation, you must be the inventor of the wheel.

Say's Law

Whether you want to be an entrepreneur or an employee that works for someone else, you must produce more value than a robot or you will be replaced by automation. Before we discuss the creative process of crafting value through products and services, let us begin by defining an economic principle that most people are not familiar with.

People, especially employees, often find it hard to understand the source of wealth creation. Employees want a raise because their cost of living is increasing; employers give raises to compensate for productivity improvements. In other words, you want a raise because your rent went up; your employer will give you a $1 per hour raise if your efforts create an additional $10 per hour profit for the company.

So how do *you* create wealth? It is not as straightforward as one would think, otherwise more people would be rich. Consider the economic principle of supply and demand from an elementary point of view. If you create a supply of items that are in demand, you will create wealth. That is the way that most people think, but not the entrepreneur.

This is an important concept, because focusing on demand always favors the robot. Humans will always lose to automation. The human wins when she creates a new supply to satisfy a previously unknown demand.

Sound backwards? To most people, it does. That is why they are not increasing their net worth, and why they will focus on developing the wrong skills needed to compete against automation.

Before we can move on to the creative process, you must first understand the principle of Say's Law, which states, "Production is the source of demand."

What? How can you have production if there is no demand? It sounds irrational, but that is exactly how entrepreneurs think. Henry Ford did not make faster horses, he made cars. Steve Jobs did not make computers, he made Macs, iPods, iPads, iPhones, and iTunes. I am not a music lover; I listen to news. I never knew I needed a music player until Jobs introduced the iPod. He created demand by producing a product I did not know I wanted.

Think of it this way. If hard work and determination created wealth, then many of those in even the poorest of communities would be rich. If hunger and demand for food created supply, then those same communities would be full of farms and restaurants. It is not demand that creates production, it is the other way around. And it is counterintuitive.

OPPORTUNITY COST

Branching out on your own as an entrepreneur or free-lancer is a scary proposition. Giving up the security of a corporate income is not for the faint of heart. A major hurdle for most people to overcome is the opportunity cost that results from forgoing a salary while building their own business. A successful wealth-building strategy is to focus on long-term goals. In this case, that would be the goal of accumulating total net worth rather than income. Let's assume your total corporate compensation package is $100,000 per year. If it takes you three years to replace this income, your startup opportunity cost could be as much as $300,000. Is it worth it? Think in terms of net worth, not income. Had you kept your corporate job, you would have been earning a salary but not building any equity. As an entrepreneur, you initially forfeited an income; however, at the end of three years, your enterprise is producing a profit in excess of $100,000 (the compensation you are paying yourself). At the meagerest of valuations, your business is worth at least $300,000 (three times earnings). So, at the end of your third year in business, you have not only a $100,000 compensation but also $300,000 in equity. You are ahead of the game and as long as you continue to grow profits, it is all icing on the cake.

This chapter introduced the first of four cognitive instructions that will be presented over the course of the book. While each can stand on its own merits, they will work best when harmonized together. *Think like a human* was introduced first because it is the keystone that holds the entire thought process together. The following five chapters relate directly to its premise, first establishing the threat of automation, and then building up to the ultimate task of developing human touch.

Action Plan

Imagine you are a safe-cracker, trying to determine the combination to a locked safe. You slowly turn the safe's dial, intently listening to hear each tumbler fall into place.

The dial you are turning is the eleven traits discussed in this chapter. As you scroll through the list, listen intently to see which traits energize your imagination. Those that call out will be the tumblers falling into place, unlocking your inner potential and aligning your personal compass with the major trends shaping the future economy.

My leading traits are:

_____

_____

_____

My supporting traits are:

_____

_____

_____

Chapter 2

THE THREAT OF AUTOMATION

I grew up in the suburbs of Pittsburgh, Pennsylvania, during the 1970s, and witnessed firsthand the closing of steel mills and coal mines that had been the source of a robust economy for nearly 100 years. The once-prosperous manufacturing regions of the Northeast and Midwest were in decline and dubbed the Rust Belt. Factories were closing and manufacturing was moving overseas. At that time, the competitive threats were primarily coming from Japan, Germany, Taiwan, and South Korea. This were before China and today's other emerging markets were such a large part of global trade.

The post-World War II booming economic era for the working class was coming to an end. Unemployment was rising in the manufacturing sectors that had previously thrived in large metropolitan areas, from Baltimore and New York City in the north and to Chicago in the Midwest. In the heart of the Rust Belt, unemployment was an epidemic. Cities that had been supported by thriving manufacturing-based economies were in decline. Young, mobile men and women were fleeing places like Pittsburgh, Buffalo, Cleveland, Cincinnati, Detroit, and Milwaukee.

Laid-off workers were collecting unemployment and waiting for the mills to reopen. In spite of strong unions and a lot of political rhetoric, the majority of factories never reopened. The young were leaving in droves, and many unemployed men in their fifties would never return to the workplace.

Although globalization took the blame for the transfer of jobs away from US companies, the real culprit was technology and automation. Yes, in some countries, labor was cheaper, but it is also true that productivity was killing labor-intensive jobs. In all years except during recessions, US manufacturing output increases. Yet manufacturing jobs have been in a steady decline since reaching a peak of nearly 19.5 million workers in 1979.

Job-killing technological efficiency comes in many forms. Sometimes it is the literal robot that replaces an automobile assembly line worker. The image of a robot makes for captivating headlines, and so that is the storyline most often hyped by the media. But more often than not, the robot is unseen like the fuel injection or the electronic ignition that makes your car drive so well. Unnoticed by most of us, the devices we use contain an army of robotic servants that eliminate jobs that had previously been done by America's working class.

In the 1970s, most cars needed regular maintenance to run properly. The common phrase for the maintenance is a "tune-up." The procedure needed to be done about every 30,000 miles, and consisted of adjusting the engine's timing and replacing worn parts like spark plugs, ignition points, and condensers. It was not overly complicated, but did require a basic level of mechanical skill that could be acquired in a high school shop class. Literally tens of thousands of men earned a living performing these simple mechanical tasks.

Today, cars do not need spark plugs changed for at least 100,000 miles, nor do they have ignition points or condensers that need replacing. There are thousands of parallel examples of how the working class earned a living performing relatively simple manual tasks that do not exist today, from unskilled physically demanding labor jobs to solidly middle income work like fixing radios, televisions, and all sorts of household appliances.

Opportunities for these workers, most of whom are men with a high school education or less, no longer exist. And it will only get worse. Products can be mass produced in robotic factories with such precision that reliability rates do not require extensive repair networks. Additionally, these automated factories can produce products at such a low cost that repair and maintenance are not economically reasonable, making it more cost effective to purchase a new product than to repair an old one.

Male job participation rates have been in decline since 1954. In the 1960s, the unemployment rate among men ages 25 to 55 was near zero. Nearly every able-bodied man had a job. Employment trends started to shift as women entered the workforce in greater numbers and as manufacturing jobs failed to recover following each successive recession. In 2016, about 12 percent of men ages 25 to 55 were unemployed for longer than 12 months. Most of these men were not actively seeking employment and thus categorized by the Bureau of Labor Statistics as "long-term unemployed," a condition that precedes permanent job lose. That represents about 10 million able-bodied men who statistically dropped out of the workforce.

Making America Great with Robots

In 2016, Donald Trump rose to prominence by promising to make America great again. His populist message of limiting immigration and restoring manufacturing jobs appealed to many in the middle class, especially those in the Rust Belt region who have seen manufacturing jobs decline for nearly 40 years. Trump was the only Republican presidential candidate since Ronald Reagan to sweep the Democratic working class stronghold states of Pennsylvania, Wisconsin, and Michigan.

Will Trump be able to deliver on his campaign promise to bring manufacturing jobs back to America? Perhaps, but if and when those jobs return to US soil, they will be performed by robots, not human labor. In 2016, the cost of an automobile assembly line spot-welding robot was $8 per hour; a US autoworker was $25 per hour. If you were paying for it, who would you hire?

A target of wrath for Trump and anti-globalists is the Trans-Pacific Partnership (TPP) trade agreement, which proposes to facilitate trade between a dozen countries. A popular media narrative of the situation portrays two US footwear companies on either side of the issue: pro-TPP Nike against anti-TPP New Balance. All of Nike's products are made off-shore, primarily in low-cost Asian countries where nearly all the work is still done by hand.

New Balance is an outlier in the footwear industry—25 percent of their shoes are made in New England factories. Their US-based manufacturing is possible because New Balance has been an early adopter of automation. By some estimates, their modern US facilities use 30 percent less human labor. Adidas, a German company unaffected by the proposed TPP, has long been pursuing the use of advanced manufacturing techniques to bring automation to the labor-intensive footwear industry. Adidas has developed a robotic manufacturing process that can greatly improve productivity. They estimate that by automating the majority of the manufacturing process, the time to make a pair of athletic shoes can be cut from weeks to hours.

Like all sectors of the economy, some of the most innovative ideas driving automation are coming from small technology startups. San Diego–based Feetz is establishing a niche with custom footwear priced around $200. Their casual shoe design is made from 100 percent recycled materials and is customizable in appearance, as well as size. They can achieve this level of customization because their manufacturing process is based on 3-D printing.

While a $200 Feetz shoe will not fit into everyone's budget, the point is that small-scale 3-D printing–based manufacturing is commercially viable in a low-cost labor sector like footwear. The trend of using 3-D printing and other forms of automation will continue to grow as cost comes down and capability improves.

As you will read in later chapters of this book, automation is not just the foe of the working class. The robots are coming for the jobs of white collar professionals. In fact, since many US-based labor jobs have already been lost to automation, proportionally, white collar jobs will be hardest hit. No occupation will be immune from the carnage of automation.

The Broader Economy

Automation-based unemployment will not be limited to developed economies. Low-cost labor-intensive markets, like those found in many parts of the developing world, will see their economies decimated. Just as manufacturing jobs left the US in the 1970s and 80s for Japan and those same jobs left Japan in the 1990s and 2000s for China, today China is losing manufacturing jobs to lower-cost markets like Vietnam and Sri Lanka. Soon, many of those same manufacturing jobs will return to robotic factories in North America and Europe.

The shift from human labor to automation will have a profound impact on the near future. Exactly when it will occur and to what extent is anyone's guess. I do not think it is productive to speculate on whether 25 or 50 percent of jobs will be eliminated, or whether the decline will start in 10 years or 25. Regardless of what may occur in the future, the threat to human labor from automation is a clear and present danger.

Over the past 60 years, male-dominated jobs, like manufacturing and construction, have shrunk from 40 percent of the market to 13 percent. This epic transformation has occurred where automation has been the easiest to implement. A similar transition is happening to the broader economy as higher-level jobs succumb to automation.

In 2016, after eight full years of recovery from the Great Recession, US gross domestic product (GDP) is averaging a historic low of 2.1 percent annual growth.[1] According to World Bank data, global GDP is barely

---

1 Eric Morath, "Seven Years Later, Recovery Remains the Weakest of the Post-World War II Era," *Wall Street Journal*, July 29, 2016, http://blogs.wsj.com/economics/2016/07/29/seven-years-later-recovery-remains-the-weakest-of-the-post-world-war-ii-era/.

growing at 3 percent. There are presently over 95 million people[2] not participating in the US job market. That translates to 36 percent of adults not looking for employment. The robots are coming to fill these jobs. Are they coming for yours?

Technology Advances in Cycles

Unlike gloom and doom prognosticators, I do not hate or fear technology. I embrace it and use it to improve my quality of life. At the same time, I acknowledge the reality that many people's lives will be devastated because their jobs will be replaced by automation. The negative effects will be felt far and wide because mass unemployment will not be limited to specific geographic locations.

Workers in Dallas or Tokyo did not suffer the negative aspects of epidemic unemployment that plagued the Rust Belt region of the US in the 1970s and 80s, but robot-induced mass unemployment will displace employees in all nations and all income levels. Advanced automatic manufacturing techniques, like 3-D printing and computerized milling machines, will make workers in Shanghai and Seattle redundant. Supercomputing expert systems will lessen the need for professionals like doctors and lawyers. Enterprise software will replace middle management. The more predictable and routine the job function, the higher the probability it will be automated, like diagnosing strep throat or searching legal briefs. High-paying professions will be hardest hit because their replacement by technology will offer the best return on investment.

Optimists will point to the fact that, as with all technological adoption cycles, the forces of creative destruction will be in play. Jobs and industries made obsolete will be replaced with new opportunities. While that is true, what the optimists are ignoring is the mean time needed to recover. Of course, the economy will balance—that is the nature of markets. What will be so painfully devastating to the unemployed is that recovery and

---

2 Jeff Cox, "What 'Are So Many of Them Doing?' 95 Million Not in US Labor Force," *CNBC*, December 2, 2016, http://www.cnbc.com/2016/12/02/95-million-american-workers-not-in-us-labor-force.html.

rebalance will likely take more than a generation; perhaps much longer, considering the magnitude of such a widespread secular trend.

History can be our guide. The Great Depression hit the US economy when unemployment rates started to rise in 1929 and reached a peak in 1933 at 25 percent. Unemployment rates stayed above 10 percent until 1941, when the US entered World War II. The decrease in unemployment only occurred because millions of manufacturing jobs where suddenly created for rearmament of the war effort and millions of Americans were put into military service. By 1945, there were over 12 million men and women serving in the US armed forces. The US economy did not truly recover from the Depression until after World War II, when the US was positioned as a global manufacturer to rebuild the war-torn global economy.

When robots start displacing humans in large numbers, many sectors of the economy will rapidly crumble. When was the last time you saw a telephone booth or the army of employees that once serviced them?

Japan is an interesting example of how the technological winds can shift. In the 1980s, Japan was the greatest threat to US manufacturing. Indeed, Japanese industrialists were in an expansionist mode, flaunting their capital and influence all over the world. Their arrogance grew as the Nikkei stock market bubble inflated. Japanese investors overpaid for prestigious US real estate like Rockefeller Center and Pebble Beach.

In the early 1990s, when I visited old decrepit factories in the Midwest, they were abuzz trying to adopt Japanese quality practices. Instead of investing in expensive automated high-speed precision equipment that would improve productivity, management adopted less expensive Japanese management techniques, like inventory control and quality meetings. Within a decade, most of those obscure factories scattered throughout the Midwest were closed.

As fate would have it, Japan did not fare any better. The Nikkei peaked on December 29, 1989, and has still not recovered nearly 30 years later. During that time, the Japanese economy has consistently been in and out of recession. Despite their superior technology and quality, Japan's fortunes collapsed. Nations like China, with lower labor costs and easy access to technology, prospered.

Now, China's reign as the world's factory is beginning to wane. I expect their economy to fall harder than Japan's. China's manufacturing edge over the past 25 years has been low-cost labor and nonexistent environmental standards. Automation and cleaner technology are rapidly eroding those advantages. The global tectonic plates of manufacturing are again shifting; automation will make low-cost labor redundant.

Near-Free Labor

Abundant automation translates to essentially free labor. This is bad news for the laborer, but good news for the consumer. Near-free labor is a hard concept for most people to grasp, just like near-free information was not understood in the early days of the information age. Prior to the Internet, just two decades ago, information was very expensive. Consequently, the best access to information was available to those who lived in large metropolitan areas or near a research university.

Most of us have forgotten or have never known how expensive information was. Even simple voice communication that we all take for granted was very expensive. In the 1960s, a long-distance telephone call from New York City to Los Angeles could easily cost several dollars per minute. When I made my first stock trade in the early 1980s, the transaction fee was over $100, something that would cost less than $10 today.

As a result of technological advances, today, information is near free. Sure, there are still subscription services for the *New York Times* (NYT), and universities still charge extortionary tuition rates. But technology is whittling away at even those last bastions of fee-based knowledge. The NYT's stock peaked in 2002, and 14 years later its price had declined almost 70 percent. Near-free information is good news for consumers, but not so much for the old information establishment.

Near-free labor will work in a similar manner. As mentioned previously, a robotic welder costs an automobile manufacturer $8 per hour while a human costs $25. A robot is not free, but the cost differential is close to 70 percent. As with all new technologies, that cost gap will continue to expand as the price of automation continues to decline. At some point in the near future, labor like information will be near free.

Combine near-free labor with the vast amounts of low-cost energy available in the US, and the result is the manufacturing renaissance that is about to take place in North America. The cost and availability of natural gas, as a result of technological advances in recovery methods from shale rock, is a case in point. Natural gas is now an abundant resource in the US, costing at least 80 percent less than in Asia. Technology has not only lowered the cost of natural gas recovery but also liquefaction, making it cost competitive to export US liquefied natural gas (LNG) in tanker ships. In 2016, the US was a net exporter of LNG for the first time since 1957.

LNG exportation is good, but local use is even better. International companies are investing billions of dollars constructing plants in the US to take advantage of low-cost natural gas. The long-term future is likely to include massive automated factories in North America (US, Canada, and Mexico) that vertically integrate the entire supply chain from energy and commodity input to finished product output: a system that will be much more efficient than today's Asia-centric manufacturing hubs.

Today this is how Asia-centric manufacturing works:

- Raw materials like coal, timber, and ore are exported from North America to China.

- China converts the imported raw materials into products that are shipped back to North America (the world's largest market).

- Waste and scrap are collected from the used products in North America and exported back to China for recycling.

This is how it will work in the automated future:

- Raw materials will be mined in North America.

- The materials will be shipped to the closest North American energy region or pipeline point, such as:

  - Natural gas shale regions, like Texas's Barnett or Pennsylvania's Marcellus.

  - Pipelines predominantly running through the Midwest, like the North/South Corridor from North Dakota to Texas, or the West/East Corridor from Utah to NY/NJ.

- Automated factories will be located along the natural gas corridors.

- Finished products will be shipped to population centers on the coasts.

Sound impossible? It will occur faster than you think. When I was a kid, my in-home entertainment options consisted of watching a black and white television with four channels: ABC, CBS, NBC, and Public Television. I had to choose between *The Beverly Hillbillies* or *Mr. Rogers' Neighborhood*. What kind of choices do kids have today?

CHINA'S RISE & FALL

Technological revolutions occur rapidly. The automation revolution will proceed unimpeded, just as the information and industrial revolutions that came before it. Economic dynamics shift quickly to adopt the new technology. China entered the twenty-first century as the sixth largest global economy with a GDP of barely $1 trillion. In 2016, China was the second largest economy with a GDP approaching $11 trillion. However, their explosive double-digit growth rate has been slowing to now barely 6 percent per year. Following in Japan's postindustrial footprint, China may be entering a 25-year recessionary cycle.

The US is about to enter a deflationary cycle not seen since the Gilded Age of the late nineteenth century. The decreasing cost to produce goods will be countered by inflationary government money printing required to compensate the unemployed. How quickly will it occur, and will automation-induced deflation outpace inflationary currency debasement? I have no idea, and neither does anyone else. I assume the transition will occur in waves, some more pronounced than others.

The deflationary forces will occur because as labor costs decrease, prices for goods and services will decline. The robots are coming to make everything cheaper. Productivity will improve in every sector of the economy, from manufacturing to medicine to national defense. The Air Force will replace $50 million fighter jets with $1 million unmanned drones. Fighter pilots that cost several million dollars to train will be replaced with ground-based drone operators that can be trained in a few months for a little more than $100,000.

Guaranteed Minimum Income

Cost savings will be abundant, and they will be needed to fund massive social spending to support the unemployed. To prevent social unrest, some form of guaranteed minimum income (GMI) will be instituted. GMI will be affordable because of robotic deflationary forces that will make the real cost of everything cheaper. Food, healthcare, shelter, and entertainment will all be less expensive, thus more easy to provide to the masses.

The concept of an affordable GMI is not as farfetched as it might seem. Since at least the Industrial Revolution, quality of life has improved at decreased cost. The average American today enjoys access to things unavailable to the super-rich just 100 years ago—simple things like air conditioning, air travel, and penicillin, just to name a few. Also, the lifestyle spread between the average person today and the contemporary super-rich is not as extreme as the concentration of wealth would suggest. An average American has access to most of the same comforts that Facebook founder Mark Zuckerberg has. Middle-income homes are not palatial mansions like Zuckerberg's, but they are warm in the winter, cool in the summer, and have a refrigerator stocked with food.

Technology has made our affordable lifestyles possible. When automation reduces the cost of labor to near zero, the availability of products and services will be even more abundant. In the 1930s, one farmer was needed to produce enough food to feed four families. Farming was incredibly inefficient, not much more productive than in feudal times. Today, one farmer's labor can feed more than 150 families.

LESS FARMERS, LESS DOCTORS

In the 1930s, farming was a solid middle-income profession. Some 20 percent of the population was employed in agriculture. Because of automation and advanced farming techniques, farmers now make up about 2 percent of the workforce. Contrast that number with today's healthcare workers, which comprise about 10 percent of the workforce. Demand for healthcare providers is growing, and they receive above-average compensation.

Like farming, much of healthcare work is routine and can be automated. Because of the rising cost and demand for healthcare, there is financial incentive to develop automated medical devices and procedures. Automation of medicine will follow the same productivity curve as agriculture's, whereby more services can be provided by fewer employees. Declining medical costs will be good news for society, but perhaps not so good for some highly compensated medical professionals whose work is replaced by automation.

The resulting abundance of food production has lowered the cost of nutrition to the extent where nearly 20 percent of US households receive direct food subsidies in the form of food stamps. If food production took 20 percent of the labor force to feed the population as it did in 1930, then agricultural costs would be too high for 20 percent of households' nutrition requirements to be subsided with tax dollars. The bottom line is that government transfer payments only exist if society can afford to pay for them.

As technology and automation reduce the cost of basic living requirements, more people's lifestyles can be funded with redistribution of wealth through tax policy. This will likely occur because automation will eliminate old labor-intensive routine jobs faster than new employment opportunities will be created. The growing number of unemployed will require living subsidies. Because automation will reduce the relative cost of basic living expenses, society will be able to fund them through tax policy.

Eventually, market forces will reach an automation-employment equilibrium. However, because of the magnitude of change and rapid implementation of robotics, I believe equilibrium will be a long time in coming—at least one generation. But, eventually, forces of creative destruction will create new employment opportunities in yet unknown professions, or the population will adjust downward to balance with available employment positions.

The average age of a US farmer is 58; the average age of a Silicon Valley technology worker is under 30. People eventually migrate to new jobs and new geographic areas. But until an equilibrium can be reached, some type of GMI program will likely be implemented to stabilize social pressures of vast unemployment. Like all aspects of automation's effect on humans, this will be received as both good and bad news, depending on one's personal situation.

Rise of the Silicon Sultans

Currently, some 50 percent of the US population receives some type of government transfer payment: Social Security, Disability, Veteran's Benefits, Medicare, Medicaid, or housing and food allowance. While GMI will meet minimum US quality of life standards and act to stabilize social unrest from the effects of massive unemployment, based on past human experience, it is unlikely to lead to a tranquil lifestyle. In communities where government transfer payments are the sole source of income, people tend to suffer higher rates of poor physical and mental health. Residents of highly subsidized communities experience much greater rates of diabetes, depression, substance abuse, and suicide than the general US population. So, while GMI may prevent social unrest, a self-actualized community is unlikely to emerge from people sustained on high fructose foods and virtual reality entertainment.

REINVENTION

Rust Belt cities like Pittsburgh lost out to Japanese industry in the 1970s. For the past 25 years, Japan's economy has been in a virtual perpetual recession. Are "has-been" cities like Pittsburgh and countries like Japan destined to collapse into an economic death spiral? No, not as long as their people return to the roots of their original prosperity—determination and creativity.

Pittsburgh's economy is not on par with the technologically advanced Silicon Valley; however, the people of Pittsburgh are no slackers. It has taken them more than a generation to recover, but Pittsburgh is shaking off the Rust Belt shackles. Pittsburgh's renaissance is built on the institutions that survived the collapse of the steel industry. Big steel did not function in a vacuum; it was supported by an ecosystem of not only mines and railroads but also universities, banks, and hospitals.

While the mines and rails have not fared well, Pittsburgh's strength has emerged by developing world-class universities, hospitals, and financial institutions. Uber's selection of Pittsburgh as the test ground for a fleet of autonomous vehicles is evidence of the city's technological savvy. Pittsburgh is home to Carnegie Mellon University (CMU), world renowned for its automation and robotics research.

Like Pittsburgh, Japan will eventually rise above its economic stagnation. In fact, I expect Japan to fare much better than China or India. Ironically, some of the issues that have plagued Japan's economy will ultimately buttress it. Among other things, Japan's aging population suffers from lack of domestic consumption, largely as a result of declining birthrates and non-immigration policies.

Declining population will likely prove to be an advantage as automation replaces the need for human workers. Japan's homogeneous, highly adaptive, and intelligent citizens will be willing beneficiaries of an advanced robotic society.

During periods of extreme technological shifts in the economy, income and wealth disparity always increases. The railroad, electric, steel, and coal industries that enriched the robber barons of the nineteenth century concentrated wealth in the hands of a few: Carnegie, Frick, Mellon, Morgan, Rockefeller, and Vanderbilt come to mind. Parallels can be drawn to the concentrated wealth of today's technology billionaires that have been dubbed the Silicon Sultans: Ballmer, Bezos, Gates, Jobs, Page, and Zuckerberg, to name a few.

In addition to the super-rich technological titans, the other winners of the robotic future will be those few people employed by large corporations or self-employed entrepreneurs. The elite, the corporate employed, and the entrepreneurs will make up the trinity of achievers. Sadly, I believe their numbers will be relatively few. Until the economy reaches an automation equilibrium, the vast majority of people will be under- or unemployed, falling into the GMI category.

The purpose of this book is to help those that aspire to be entrepreneurs or corporate survivors. While the lack of extraordinary talent and luck prevent most of us from achieving the extreme wealth of the upper 1 percent, a financially independent lifestyle is within reach of those with determination and discipline. In the following chapters of this book, I will present information and strategies to help you prepare for the inevitable robotic onslaught. The successful will learn to implement technology, not fight it.

Action Plan

Now it is time to reevaluate your skills in light of the threat of automation.

1. My three unique critical success skills are:

    a. _____

    b. _____

    c. _____

2. My skills can/cannot be easily automated (identify the technologies than can be used and the degree to which automation can be achieved):

    a. _____

    b. _____

    c. _____

3. For skills that can be easily automated:

    a. These are commodity tasks and do not provide me with a competitive advantage.

    b. Identify new skills that will make me uniquely positioned to achieve my stated goal.

4. For skills that cannot be easily automated:

    a. They will act as protective moats and reduce the chances that my abilities will be replaced by a robot.

    b. Pursue and improve these skills.

Chapter 3

ENABLING THE DISADVANTAGED

There is much to be concerned about in regard to the threat of automation displacing workers and creating a catastrophic shift in the economy. However, mankind is a resilient and adaptable creature that, throughout history, has mastered his surroundings, from the deserts of the Middle East to the icecaps of Antarctica. Conquering the impending threats of automation will be achievable for the few that prepare and take action.

The course will be complex, yet easy for the individual who constructs a path that slopes with the technological trends. Interestingly, those who have been disadvantaged are most likely to adopt technology early on to gain an edge over a comparative handicap. To paraphrase an old proverb, "the meek shall inherit the earth."

The term "disadvantaged" is relative. If you suffer from diabetes, hypertension, or epilepsy, you probably do not consider yourself disadvantaged or disabled. These conditions are relatively inconsequential if kept in check with medication. However, someone with the same illness 100 years ago would likely have died before middle age. Today, these maladies are an inconvenience.

To emphasize technology's enabling effect on the human condition, look back in history to the era of hunter-gatherer societies. Something as simple as nearsightedness would have prevented a man from becoming a skilled hunter and thus a contributing member of the clan. He would have been relegated to performing less productive tasks, thus limiting his usefulness and likely his social stature. Despite the fact that he had a keen intellect or possessed strong physical ability, poor vision may have prevented him from finding a mate and bearing offspring. A simple visual impairment could have ended his genetic future, relegating his family line extinct.

Corrective lenses and Lasik eye surgery have relegated nearsightedness to merely an inconvenience. A person with a corrected visual impairment is relatively unconstrained. For example, this person could have a career as a pilot, something that would not have been the case a few decades ago. What will the technology of tomorrow disqualify as a disability?

Today, someone that is blind may be hindered by a similar stigma as the Paleolithic nearsighted hunter-gatherer. Imagine a person that has the genetic artistic talent of Rembrandt, yet is physically blind. Today, that person would be prohibited from reaching their potential as a visual artist. In the future, technology will be developed that will allow the brain's neurons to visualize one's surroundings, even without physical sight. A blind woman will be able to not only visually sense her environment, she'll be able to translate that expression into a visual image for all to enjoy.

Something as simple as nearsightedness would have put an undue strain on a primitive hunter-gatherer society 10,000 years ago. Today, people with complex disabilities are able to live fulfilling lives and contribute to society. At some point in the not-too-distant future, medical devices and implants will mitigate the disabling effects of most physical handicaps.

The consequences of a disability or disease are often conquered as much by applied technology as by pure medicine. A few examples from history are noteworthy. Consider the conditions of Helen Keller and Stephen Hawking.

History Is on the Side of the Disadvantaged

Shortly after Helen Keller was born in 1880, she contracted a serious illness that left her blind and deaf. Technologies like Braille and teaching methods such as elocution were being adopted to teach the blind and deaf; however, information was not readily available in the 1880s. This was especially true for the Keller family, who lived in the rural state of Alabama. It was somewhat serendipitous that her mother learned of an educational program for blind and deaf children from a Charles Dickens story that had been published some 40 years prior. Imagine living in a world where it took decades to disseminate "new" information and access occurred through happenstance.

The Keller family's persistence in helping Helen was rewarded with a chance meeting with the inventor Alexander Graham Bell. Bell, a wealthy industrialist, was the Steve Jobs or Bill Gates of his day. Bell had a long history of working with the deaf. His mother had lost her hearing and his father had invented a visual speech technique that was used to teach deaf children to communicate. Bell's experience educating the deaf led him to a fascination with the newly emerging science of sound. His study of sound eventually led to his invention of the telephone and phonograph.

Through the association with Bell, Anne Sullivan was hired as Helen Keller's personal tutor. Sullivan was uniquely qualified to teach the insular and unruly seven year old. Sullivan was a pioneer of applying emerging technologies to teach the blind and deaf. The sign language methods that she taught Helen had not yet become standardized into today's American Sign Language. In addition to her technical expertise, Sullivan possessed an extraordinarily strong personality that broke through the barriers of the troubled little girl. Because Helen was both blind and deaf, she had a very limited understanding of the outer world. Extraordinary determination was needed by both pupil and teacher. The merging of their efforts is an example of what I call human touch.

Helen Keller was able to master communication with the enabling technologies of her day: sign language and Braille. She was the first blind person to earn a college degree and she went on to have an inspirational career as author, lecturer, and activist. Helen's rise above her disability was enabled by both technology and human touch.

Helen Keller died at about the time that young Stephen Hawking was diagnosed with amyotrophic lateral sclerosis. Like Keller, if not for advancing technology, Hawking would have been captive to the degenerative disease. But, coincidently, Hawking's deteriorating physical condition and loss of speech occurred in 1985, at the dawn of the information age. This was a time when relatively inexpensive consumer-level computing power and software were being developed. Small, inexpensive servo motors that were easily controlled by direct current and could be used to power wheelchairs or open doors were becoming available. Rudimentary computer-controlled voice synthesizers were developed to enhance communication. So perhaps more importantly than medicine, it is computing power and small electric motors that enable Hawking to continue his scientific work. Had he been born merely a decade earlier, it is likely that his potential as a cosmologist would have never been realized, and his gifted genius would have been confined to the prison of an invalid's body.

Consider the potential to society when physical disabilities are extinguished. How many potential Rembrandts could not paint because they were blind? Technology will enable the otherwise disadvantaged. Society will reap huge rewards when the untapped potential of every person is released.

Use Technology to Disable Your Disadvantage

To begin to visualize automation as an advantage rather than a threat, start by assessing the enabling benefits. View technology through the eyes of a disadvantaged person, someone whose life could be greatly enhanced by applying a new technology. Then consider your own physical impairments or shortcomings. Identify your weaknesses for what they are and then look for technological solutions.

Disadvantage is a relative term. It can mean a physical impairment, as in the above examples; however, from a broader perspective, it can be interpreted as any impediment, from the societal discriminatory forces of "ism" (racism, sexism, ageism) to the self-imposed limitations of discouragement or addiction.

BREAKING THROUGH THE GLASS CEILING

Women account for nearly 46 percent of the workforce yet hold less than 5 percent of Fortune 500 CEO positions. At least 30 percent of these large firms headed by women are in a technology-related sector. Technology and innovation are the great equalizer and can help women break through the corporate glass ceiling. Better yet, technology can help women create their own entrepreneurial firms, thus entirely bypassing the male-dominated corporate structure.

So, what holds you back from reaching your goals? We all have at least one impediment, whether physical, societal, or personal. Reach out to your spouse, partner, family, or friends; whoever you think will give you an honest answer.

The key to profiting from the automation revolution will be using technology as a tool to first overcome your own personal disadvantage and then to create products and services for others. Humans can never compete with robotic automation when it comes to repetitive tasks. Fortunately, profits are made by solving problems, not by completing tasks.

Gaining Your Advantage Over a Robot

1. Determine the critical impediments that are keeping you from advancing your career or business objectives. These could be complex or simplistic, personal or systemic. For example, you might have a speech impediment, lack credentials, or simply not have capital to fund your project.

2. Determine if a technology exists to solve or mitigate your weakness.

3. If a solution exists:

   a. Buy it now or plan to buy it in the future.

   b. Press forward as if the hindrance does not exist.

   c. Spend your precious resources of time and capital on developing your uniquely human talents that set you apart from others.

4. If a solution does not exist, then your goal will not be easily achieved:

   a. Scrap your goal.

   b. Select a new goal.

   c. Start again at Step 1.

This process is effective because of its elegant simplicity. It allows you to cut your losses quickly and move on to attainable solutions.

A key point to keep in mind is that the technological solution does not have to be affordable or readily available; it just has to exist or have the potential to exist in the near term. This is because technology is deflationary in nature, meaning that over time it will become more available and at a lower cost. If there is a technological solution, then your current weakness can be overcome, perhaps at a future date. This allows you to focus on your strengths rather than waste resources on your weaknesses.

Technology in Action

Here's a simplistic personal example from my own life. I have absolutely no sense of direction. None. I can drive to a destination multiple times and yet still get lost because I fail to observe landmarks or other reference points. I simply do not "see" them. It could be due to poor short-term memory, lack of spatial recognition, or most likely, I just get caught up in daydreams and simply do not pay attention.

This was a notable hindrance to my career as a traveling salesman in the early 1990s, pre-GPS and smartphone technology. As I traveled the Midwest visiting customers, I had to spend an inordinate amount of time planning my trips, especially those last few miles off the interstate on local side roads. Without detailed instructions, I would always get lost.

This affected my productivity as a salesman because I had to spend more time than the average person planning out my route. I was sure there was some method to overcoming my directional dyslexia, especially if I dedicated hours to defeating it. However, I chose to remain directionally ignorant and instead exert my resources on developing sales and business skills. So, today, I still get lost easily, but in spite of that shortcoming, I am a very capable and highly compensated businessman.

The reason I did not try to improve my directional ability was that I knew that military GPS technology would someday be available at an affordable price to the consumer civilian market. In a relatively short period of time, I knew that my "disability" would be easily solved with the aid of an inexpensive piece of electronics.

Rather than try to fix my imperfection, I purchased a solution, thus allowing me to develop more profitable unique human skills.

INVESTOR TIP

For new technology ideas, look to early adopters like the medical profession and the military. Also, keep an eye on Japan—they are manufacturers of world-class electronic devices and will be developing many robots to aid their aging population.

Be an Early Adopter of Technology

In this section, I referenced Helen Keller and gave several examples of how vision, or the lack thereof, has been overcome with technology. I chose vision as a metaphor because it is important as both an esoteric skill (the ability to see future possibilities) and a practical solution that is always being improved upon with technology.

For example, "vision" technology, known as augmented or virtual reality, exists today. On a consumer level, it is being developed for entertainment, like experiential 3-D movies or gaming; a recent example would be the app game Pokémon Go. It also has a practical application as an aid in performing critical tasks, such as surgery.

Several years ago, Google Glass received a lot of scrutiny because of its ability to surreptitiously record video. It was considered an invasion of privacy and creepily voyeuristic. The product was pulled off the market, but at some point, an improved version will emerge. Early adopters of this technology will likely be surgeons or others performing expensive and complex tasks where a visual aid will be beneficial.

Imagine a surgeon performing an operation. Looking through her augmented reality (AR) device, she not only sees the patient's body, but also other visual cues superimposed in the glass. Perhaps, she sees an outline of where the incision should be made. Or maybe she sees a step-by-step, highlighted guide of how the procedure should be conducted. Or, perhaps, there's a real-time display of an MRI revealing what is occurring out of sight within the patient's body. The possibilities for providing a surgeon with real-time visual (and audio) cues are limitless, and they arguably would improve the performance of even the most skilled doctor.

Applications for AR will advance, from the early adopter professions like medicine or the military, to everyday applications like auto mechanics.

In the 1970s, a mechanic maintained a library of Chilton Repair Manuals as a reference to provide step-by-step instructions on repairing specific brands of automobiles. In the 2000s, that data was made available on CD-ROM. Today, it is available online. In the not-too-distant future, the cost of AR devices will be affordable enough for use by an auto mechanic. He will wear a Google Glass–like AR device that will provide him with a visual and audio reference to assist with the repair of an engine.

SMALL BUSINESS ADVANTAGE

Small business entrepreneurs might consider themselves disadvantaged by large firms with multimillion-dollar budgets. To avoid this trap, a small business should avoid competing in areas that require large amounts of capital investment and instead focus on their comparative advantage; for example, nimbleness. Back in 2005, multinational corporations were purchasing $200,000 teleconferencing systems for their boardrooms. These systems were complicated and unreliable. Executives were incapable of operating the devices without the assistance of a technician. At the same time, small business entrepreneurs were rapidly adopting Skype as their preferred communication method. Skype was user friendly and either free or very inexpensive to operate. Entrepreneurs flocked to it. Corporate executives living in their information technology–cocoons did not know it existed.

Action Plan

1. My goal is to _____

2. The three critical skills that will make me successful are (hint: think in terms of your unique human talents):

   a. _____

   b. _____

   c. _____

3. The three critical weaknesses that will hinder my progress are:

   a. _____

   b. _____

   c. _____

4. I will use the following technologies to mitigate the critical obstacles:

   a. _____

   b. _____

   c. _____

5. If a solution does not exist to mitigate the obstacle:

   a. Redefine or select a new goal.

   b. Start the process over again.

6. If solutions do exist, now or in the near future:

   a. Purchase or plan to purchase the solution.

   b. Focus on improving the unique success skills in Step 2.

Chapter 4

SEEK PRACTICAL EDUCATION

Our education system has its roots in eighteenth century Prussian structure, designed to educate a conformist workforce. The system operated well for several hundred years, educating standardized cogs to fit the wheels of the Industrial Revolution. However, it began to break down near the end of the twentieth century as the information age was taking shape.

Today's millennials owe over $1 trillion in student loan debt, and real wages have been stagnant for over a generation. The system is badly in need of reform. Automation's killing of jobs will hasten the transformation.

A decentralized market-driven system will eventually replace today's antiquated educational hierarchy. This market will be driven by both demand needs of the employer and the supply of individual, unique talents. The good news is that you do not have to wait for the change to occur. Steven Spielberg was twice rejected from the University of Southern California's Cinematic Arts program. Instead, he studied at the less prestigious California State Long Beach, where he eventually dropped out. This is a classic example of how the university system either ignored or failed to appreciate the unique talents of a gifted individual. In the future, it will be unlikely that aspiring film directors will even bother to attend college.

They will hone their skill and earn accolades by posting films on direct-to-consumer sites like YouTube.

Credentials

Do you need a degree? It depends on your career objectives. If the biological sciences appeal to you and you wish to become a medical doctor or veterinarian, then, yes, formal education is required. What if you love animals but do not want to go to vet school? Look for ancillary careers. Niche markets always exist, and they can be extremely profitable. The average veterinarian's salary is about $90,000 per year. I personally know people that are self-employed in the animal control industry who are making $100,000. They humanely remove critters like raccoons and skunks from suburban neighborhoods, no formal education required.

Does computer science appeal to you, but you cannot afford to attend the Massachusetts Institute of Technology? No problem. In recent years, I have personally met several people with only high school diplomas that are employed in the Internet security industry and are earning in excess of $200,000 per year. They're not easy jobs to come by, but they do exist for those with digital talent. If you are skeptical, Google Edward Snowden, the National Security Advisor whistleblower. Snowden is rumored to have only a high school GED, no college degree, yet prior to becoming a whistleblower, he was earning something in the range of $122,000 to $200,000 per year.

My advice is to worry less about credentials and to look for opportunities that align with your interests. If education or certification are required, then by all means, find a way to attain them. Otherwise, proceed with caution and find ways to improve your overall competence in the field that appeals to you most.

Do Credentials Matter?

The current university system does a good job of issuing credentials, but not necessarily certification. Someone with a Harvard law degree is obviously intelligent because they had to first gain admittance to this prestigious institution, then they had to graduate. However, intelligence does not ensure competence in a given field.

Albert Einstein was a mathematical genius, but absentmindedness would probably disqualify him as a tax CPA. Steve Jobs was a visionary entrepreneur, but lacked social skills desired in most corporate settings.

Consider how you make hiring decisions in your own affairs. I never asked my dentist, mechanic, or CPA where they received their education. I select professionals based on references from friends or social media. "Where did you get your root canal?" "Who's an honest mechanic?"

The university system and affiliated institutions (e.g., American Medical Association, American Bar Association) have had a monopoly on issuing credentials. Social media and opportunities to bypass government regulation will diminish these monopolies.

Dental tourism in Mexico is an apt example. Dental work is expensive, and over 150 million Americans are not covered by dental insurance. Senior citizens are especially affected because of the long-term effects of degenerative dental diseases. For decades, seniors have been traveling south of the border to find less expensive alternatives. A trip to Mexico can easily save a senior over $10,000. This trend is growing exponentially because of access to information and easy coordination via the Internet. These patients do not care if their dentist has a degree from a US university or is board certified. It's not just dental procedures that are in demand; medical tourism in general is a growing trend as well.

DEMISE OF OPEC

Do you think that large institutions like the educational or medical establishments cannot be forced to undergo dramatic structural change? Consider the half-century reign of the Organization of the Petroleum Exporting Countries (OPEC). OPEC has had pricing power due primarily to the extremely large and influential production capacity, also known as swing production capacity, of Saudi Arabia. Most people believed this monopolistic stranglehold could never be broken as long as fossil fuels were a predominant source of energy. But technological advances (fracking and directional drilling) in extracting oil from shale deposits has doubled US production and put it on par with that of Saudi Arabia and Russia. The price of oil has fallen from a $145 per barrel peak in 2008 to consistently below $50 per barrel in 2016. Swing production capacity now rests in the hands of 2,000 shale oil wells dispersed throughout North America. Saudi Arabia is pumping oil at historic levels and yet running nearly $100 billion budget deficits. Peak oil has not yet materialized, but we might be witnessing peak OPEC.

Death of Lecture-Based Education

Automation and technology will make classroom lecture–style learning obsolete. Unfortunately, today's online courses simply mimic the boring lecture method of teaching and are generally just as ineffective. Learning in the future will be dynamic and interactive. Identifying the student's potential and enhancing their learning experience with virtual reality and simulation will play a much larger role.

As education methods advance beyond the vintage Prussian style, it is likely (though perhaps counterintuitive) that they will take on positive aspects of more ancient training methods. I foresee the resurgence of apprentice programs. This is a more productive method of learning a skill or trade, and it transfers the responsibility of education away from the government to the employer and individual.

As economies progress, the needs of employers become more specific. In 1950, an electrical engineer possessed the skill set to work across many industry sectors; he could easily find employment at General Electric, General Motors, or Bell Telephone. Today, a software engineer has very task-specific skills that are not readily transferable. A programmer at Microsoft may not find her coding skills relevant for employment at Google.

From both a practical and a proprietary standpoint, it is to the employer's advantage to train their own employees. This practice has been avoided in recent generations due to cost. However, the advent of virtual reality–based interactive training will bring the cost of customized education down substantially, thus incentivizing employers to develop proprietary training methods.

The platform for education will no doubt be highly dependent on technology, drawing on virtual reality, big data, and yet-nonexistent sensory devices. However, the actual structure of the educational process is likely to be a modern-day adaptation of the time-honored apprenticeship. For those not familiar with an apprenticeship, it is a form of on-the-job training that includes a time commitment from the employee (historically seven years).

This type of educational program was used throughout history because it provides a win-win outcome for both employer and employee. The system produces a skilled worker who meets the exact needs of the employer, and the cost of training is shared between both parties. The employer provides the training in exchange for work at discounted compensation over a set period of time. The employee receives training at no cost, and gains relevant work experience and some form of monetary compensation.

At the conclusion of the apprenticeship, the employee is a free agent with options to remain with the current employer, seek other employment opportunities, or start their own enterprise. Variations of this system have been used successfully for millennia. The US military has more or less used an apprentice-type system to field the all-volunteer services for more than 40 years.

Some might be worried that apprentice programs could evolve into indentured servitude. Perhaps, but compared to our current student loan dilemma, how much worse could it get?

EDUCATIONAL CAREER OPPORTUNITIES

The career outlook for the traditional college professor might be on the wane, but huge money-making opportunities will exist in education. The automation revolution will produce at least two distinct needs in education: training of displaced workers, and training (programming) of robots. Opportunities will abound for all types of experts: software and hardware engineers, psychologists, communicators, integrators, etc. The sky is the limit. Where will you fit in?

Disruptive Thought Experiment

Are your alternative learning juices flowing? Let's perform a thought experiment.

This exercise is designed for two outcomes:

1. Nudge the skeptical professional toward the possibility that they are not immune from being made redundant by automation and technology.

2. Spur creative thinking.

The thought experiment consists of brainstorming a disruptive new product or service. For our purposes, we will use a simple two-step approach:

1. Identify an opportunity that is evident from an existing market-based need or desire.

2. Develop a solution using new technology that creates value (i.e., saves money or time).

    a. The solution can be either a product or a service.

    b. The technology can be off-the-shelf or at least commercially feasible in the near future.

You will need to come up with an opportunity and solution on your own, based on your personal interest and skill set. To help you envision the process, I will walk you through an example.

Opportunity: Create a lower-cost orthodontic solution for straightening teeth.

- Market: Americans spend $11 billion dollars annually in the pursuit of aesthetically straight teeth.

- Orthodontists are highly compensated professionals, earning an average income of $200,000 per year.

Solution: In-home DIY 3-D printed retainer kit.

- Value: Saves money by bypassing the services of an orthodontist.

- Method: Obtain a bite impression.

  - Low-tech approach: Cast an impression using a traditional molding method (e.g., Polystone), and then create a 3-D scan of the mold.

  - High-tech approach: Develop a low-cost sensor that can be placed in the mouth, which would create a 3-D image.

- Develop CAD software that transitions condition of existing bite by incrementally moving teeth to the desired aesthetic position.

- 3-D print a series of plastic retainers that can be worn over time to incrementally correct the bite.

Sound too easy, or like an unrealistic blue-sky solution? After I worked through this thought experiment in my own mind, I Googled the idea. I found that an undergraduate student from the New Jersey Institute of Technology named Amos Dudley had already pioneered the concept and used a solution not too different from my own. He did not just think through a thought experiment. Dudley used off-the-shelf technology to 3-D print custom retainers that actually straightened his teeth. He spent about $60 in materials to produce a result that would have cost several thousand dollars going to an orthodontist. Dudley's exploits can be researched at: http://amosdudley.com/weblog/ortho.

Dudley is a skilled 3-D design engineer, so I am not suggesting that his process enables any old Joe Sixpack to print retainers at home. Not yet. It is proof-of-concept that such a method is commercially viable and would suggest that consumer-level solutions could be available in the not-too-distant future. If I were an orthodontist I would be concerned; very concerned.

Key Takeaway

If you are a skilled professional who is skeptical that your expertise can be replaced by automation or technology, think again. It may not occur tomorrow, but history teaches us that technology eventually advances to the state where it can produce a viable competitive alternative. DIY orthodontics may seem farfetched, but the same could have been said for now-common procedures, like in-home blood sugar monitoring, pregnancy tests, and DNA sampling.

For the rest of us, disruptive brainstorming is not an elitist exercise. Give it a try and see where it leads you!

Action Plan

Modern apprentice programs might be in the future, but you can get ahead of the learning curve by taking action now. Use this exercise to identify training opportunities from nontraditional sources.

1. Identify an employer I can apprentice with through either full- or part-time employment: _____

2. Identify manufacturing companies that will train me to use their products:

   a. _____

   b. _____

   c. _____

Identify service companies that will train me to use their services:

    a. _____

    b. _____

    c. _____

4. Identify experts in my area of interest and how I can engage them:

    a. Conferences: _____

    _____

    _____

    b. Organizations: _____

    _____

    _____

    c. Training or consulting expert: _____

    _____

    _____

5. Identify online training from free sources like YouTube or fee-based providers:

    _____

    _____

    _____

Chapter 5

TRANSFORM KNOWLEDGE INTO WISDOM

A key success factor for thriving in the automation revolution will be the ability to transform knowledge into wisdom. A human cannot be any more productive in knowledge attainment than at manual repetitive tasks. A computer-enabled robot will always win. It has infinite storage capacity and executes at the speed of light.

Humans require years of training to become competent in a complex skill and can take decades to attain expert status. In the 1950s, a machine operator would require some 4,000 hours of training to become certified as a journeyman machinist. Years of experience were required to set up and operate a machine in exactly the same way so that parts could be manufactured to a consistent standard tolerance, measured in thousandths of an inch (about the diameter of a thin human hair). Today, an inexpensive off-the-shelf computer numerical control (CNC) machine comes programmed to meet critical tolerances. Depending on the task, little to no operator training is required. A simple microprocessor can operate a three-axis milling machine at a fraction of the cost of employing a human machinist. Computer automation has made the skilled production machinist obsolete.

A similar fate awaits many professions. A pharmacist's or a barista's skill set is not that much different than that of a machinist: knowledge of materials, ability to follow directions (a formula), and precision mixing of compounds. These are all tasks that, just like the CNC machine, can be mimicked by a robot, and at higher levels of productivity and quality control. Robots operate around the clock and do not become addicted to narcotics like pharmacists, nor do they give away free coffee to their friends like baristas.

LEVERAGE YOUR KNOWLEDGE

Professionals with subject matter expertise can monetize their experience by developing decision support systems; in other words, computer and smartphone apps. The possibilities are endless, from the practical to the frivolous. Imagine the demand for a comprehensive medical diagnostic tool that could assist a physician's diagnosis of a rare disease or help a parent determine if their child's earache was worthy of a visit to the emergency room. The market is even larger for the trivial: selecting a dinner wine, matching an outfit, or buying a car.

The potential goes well beyond today's simplistic smartphone-powered apps to all-encompassing robotic systems that provide knowledge and function. Today you ask Siri for directions, tomorrow she will drive you there in an autonomous car. It was not coincidental that Apple bought Didi Chuxing, China's version of Uber.

The development and integration of inexpensive sensors, speed-controlled servo motors, wireless technology, and omniscient databases are just around the corner. The futuristic is nearly a reality.

When I was a kid, there was a flying car concept based on the Jetsons cartoon. I am still waiting for that technology, but what is here is almost as revolutionary. George Jetson's car was capable of flight, but George still had to operate it. Today, autonomous cars are a reality, from experimental fleets at Uber and Google to off-the-shelf affordable options like Subaru's EyeSight® "driver assist" technology. I drive my Subaru on the interstate for hours without having to touch the gas or brake pedals.

How can you monetize your expertise?

Expert Decision Support Systems

Robots are ubiquitous: traffic lights that control the flow of cars, ATMs that dispense cash, pacemakers that regulate heart rhythm. The difference between these robots and future robots is in the degree of intelligence, much like people.

Successful products contain just the right amount of intelligence to complete the desired task, a tradeoff between function and cost. Smoke detectors employ an ionization chamber and a simple circuit to detect the presence of smoke, not a complicated gas chromatography mass spectrometer instrument that an analytic chemist would use in a research laboratory. Maintaining human relevance will require efficient use of tools in all their forms: robotics, big data, cloud computing, etc.

Advances in artificial intelligence (AI) have occurred at an exponential rate. The next generation of AI will form networks that utilize the infinite resources of the Internet and cloud computing to make storage capacity a nonissue.

Think of this in terms of the way people once listened to recorded music. To hear a favorite song, one had to purchase the vinyl record or wait for the song to be played on a radio station. At public places like restaurants or bars, there was a jukebox that contained maybe 50 of the most popular songs. Technology existed to record music; however, access was limited by the bottleneck in the distribution system. Likewise, AI software exists today and is improving at an exponential rate; however, practical implementation will be limited until large databases containing expert knowledge can be linked across the infinite capacity of the Internet.

In the near future, complex algorithms will run expert decision support systems that tap into vast knowledge databases. Ultimately, these systems will become self-learning, thus expanding their knowledge bases without human intervention.

What is a decision support system? Think of a simple flow chart. If observation "A" exists, then function "B" is carried out for a desired outcome "C."

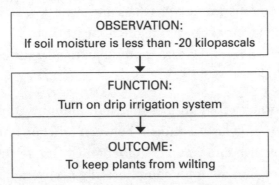

The system can be simple or complex, limited by the function of the algorithm and the factual knowledge of the database.

Consider the way a child develops decision-making skills:

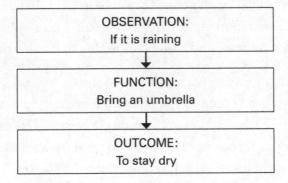

Such cognitive ability is an application of logic and differs from complex thought only in the size of the knowledge base. For example, a small child might be able to recognize the condition of rain, but not be able to differentiate between a storm and a hurricane. Likewise, a child's knowledge of acceptable functions and desired outcomes would be limited. The function and complexity of decision support systems evolve in a similar manner to how a child develops reasoning skills.

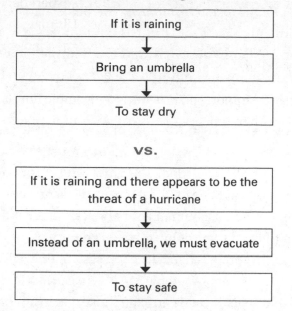

For another example, let's look at how tax preparation has evolved from professional to DIY to a nearly autonomous decision support system. In the 1960s, an accountant might be hired to prepare one's tax return. The accountant possessed knowledge of the tax code in her mind and used an electronic calculator to perform simple arithmetic. In the 1970s, the accountant would likely have replaced the calculator with a business computer that automated some of the tax calculations. By the 1990s, consumers would purchase do-it-yourself (DIY) tax preparation software to bypass the need for a professional accountant. Today, taxpayers use smartphone apps to automatically collect receipts and take suggested methods for reducing taxes on a daily basis. At each step, the AI device operated at a higher level of reasoning (calculator to business computer to consumer software to app).

AI is presently at the cusp of early-stage reasoning. The current status of decision support systems is limited by the extent of expert databases and hardware (e.g., robotic arms and sensors), not computing power.

Expert databases are being developed to compile all aspects of human knowledge. Think of an expert database as an extension of Wikipedia. Prior to the Internet, reference material was contained in hardbound books like the *Encyclopedia Britannica*. The books had to be purchased and stored on a physical shelf, printed content space was limited, an index was the only form of search, and the information was not easily updated. Wikipedia vastly improved the access of general information compared to using a hard copy encyclopedia.

Today we access medical knowledge through a physician. If you have a pain in your chest, you visit the emergency room to determine if it is a heart attack or just indigestion. The emergency room doctor's mind functions as your portal to a medical expert decision support system. It is a system that requires a human with above-average intelligence and decades of training. At some point, this vast amount of knowledge will be transformed into a computerized decision system that will be accessible wirelessly via the Internet. Wearable sensors (e.g., wrist straps, implants) will monitor vital statistics and correlate those observations with desired medical outcomes.

To some degree, these systems already exist, but they are primitive compared to what will be commercially available in the next decade. Wearable sensors will be ubiquitous, enhancing all aspects of our lives. A jogger experiencing chest pains would receive an alert instructing her to remain calm, that she is having a heart attack and an ambulance is on the way.

NO EXPERTISE...NO PROBLEM

Feeling left out because you do not have the knowledge to create an expert system? The good news is that you will greatly benefit from the creation of such technology, similar to how nimrods like myself benefit from spell-check, TurboTax, and innumerable other software solutions.

The difference is that unlike current software, functional expert systems will be capable of providing both services and products. Expert systems will provide the layman with affordable solutions to complex problems—legal, medical, mechanic, and trivial. Today, you ask Alexa to ship an item from Amazon. Tomorrow, she will determine your need and then instantaneously 3-D print the product solution in your home.

A Robot Is a Tool

When all knowledge is correlated into expert decision support systems, what need will there be for human expertise? Humans will only become irrelevant when they cease to be human. Journeyman machinists are no longer needed to mass produce precision parts; however, humans are still employed to design and create new products. A robot is a tool used to efficiently manufacture an item; a human is needed to conceive of the item.

Think back to Henry Ford's comment about his customers wanting faster horses. IBM introduced the first smartphone in 1994, called the Simon Personal Communicator. More than a decade later, mobile phones still did not possess much intelligence. In 2007, the most predominant physical feature of a mobile phone was its keyboard. You have probably forgotten, but Blackberry's keyboard technology was a major competitive advantage. That year, Apple launched the iPhone, which replaced the keyboard with a large functional touch screen. The lack of a keyboard was the focus of much critique, both positive and negative. The touchscreen was transformational in developing a smartphone with real intelligence. Steve Jobs did not simply manufacture a better mobile phone with incremental improvements, like a better keyboard. His visionary genius created an entirely new platform for managing communication, information, entertainment, and much more.

Virtual Robot

In addition to thinking of a robot as a tool, it is important to remember that a robot is not necessarily a physical object. When thinking about robots, think in a broad sense: Robot is simply a term for automation of a task. A robot could be a physical device that vacuums your carpet or an alarm system that monitors the security of your home. A robot vacuum cleaner replaces a human maid. A security system replaces a human security guard. Spreadsheets replace manually entering numbers into a calculator or cyphering mathematics in one's mind.

THE ROBOTS ARE ALREADY HERE

Smartphones are robots that have replaced other products: GPS, camera, recorder, calendar, flashlight, the list goes on. The effect has been devastating on consumer electronics industry brands like Panasonic, Canon, Garmin, Casio, and Nikon. As a consumer, you are extremely happy about this, much like your employer will be when he replaces you with an inexpensive robot. Many companies are struggling for relevance in the age of the smartphone. Workers made redundant by automation will suffer a similar fate.

A critical threat to human professionals and institutions is the sheer simplicity and amorphous form that automation can take. Consider the relevance of the Federal Reserve. The Federal Reserve was established in 1913, and with each passing year, it has increased its scope and authority over the economy. As a consequence of the 2008 financial crisis, the Federal Reserve expanded its balance sheet to over $4 trillion, approximately the market capitalization of all 30 companies on the Dow Jones Industrial Average. Wall Street pundits analyze every word emanating from the Federal Reserve in an effort to predict interest rate movement. The Federal Reserve's influence over the global economy has never been greater. Yet this powerful institution's authority is being undermined by a simple algorithm: Bitcoin.

During the depths of the 2008 financial crisis, an article was written describing the concept of a peer-to-peer electronic cash system that used blockchain technology. Think of the blockchain as an open public ledger where the record cannot be altered. The data is stored on thousands of computers across the globe and is instantly accessible. Once a transaction has been recorded, it is permanent. It cannot be censored or manipulated by any individual, corporation, or government. As an undisputed digital depository of record, it could ultimately replace long-established institutions, like county court houses, real estate title companies, patent offices, etc. The possibilities are endless.

The most popular use of blockchain technology is the cryptocurrency named Bitcoin. What started out as an anarchist novelty during the depths of the Great Recession has become an omnipotent medium of exchange. By 2016, there was over 15 million in Bitcoin circulating, with a market capitalization exceeding $9.5 billion. In less than a decade, Bitcoin's value had exceeded that of 113-year-old American icon Harley-Davidson. Harley-Davidson manufactures nearly 300,000 motorcycles per year, employs over than 6,000 people, and has assets of $4 billion. Bitcoin manufactures nothing, creates no revenue, has no employees, and owns no assets. Bitcoin is a virtual, robotic central bank. It exists simply as a blockchain database algorithm run on a network of unaffiliated computers.

Bitcoin is relevant because it has perceived value. That value is determined by the interaction of millions of humans, not the monetary policy of a central bank. Bitcoin or similar cryptocurrencies using blockchain technology could eliminate the need for central banks and treasury departments just as Napster file-sharing technology made the music industry of the 1990s obsolete.

HUMAN ARROGANCE

Misapplied automation will expose human folly and thus generate opportunity for those with real wisdom. Following the 2008 financial crisis, former Federal Reserve chairman Alan Greenspan made a startling admission about the failure of the Federal Reserve to predict the financial crisis: "It all fell apart, in the sense that not a single major forecaster of note or institution caught it." He added, "The Federal

Reserve has got the most elaborate econometric model, which incorporates all the newfangled models of how the world works—and it missed it completely."

Greenspan's acknowledgment is particularly noteworthy because for 18 years, he presided over the Federal Reserve and was dubbed "Maestro" for his orchestration of the economy. His admission of failure is a real-world example of the computer science euphemism "garbage in, garbage out," meaning that poor quality input will always produce faulty output. The Federal Reserve's computer models were no doubt based on academic intelligence, but lacked wisdom.

Artificial Intelligence Versus Human Wisdom

Artificial intelligence is not called artificial *wisdom* for a reason. Intelligence is the ability to apply knowledge for a desired outcome. Wisdom is more nuanced and perhaps esoteric. It is the confluence of knowledge guided by the human experience. I stress human, rather than experience, because robots are able to learn through trial and error. Wisdom acquisition is a product of both logic and emotion.

Intelligence tells us that two plus two is equal to four. Wisdom tells us who we should marry or what career we should pursue to optimize happiness. I believe that intelligence is the discovery of natural laws (i.e., physics) and wisdom is the application of those laws for self-actualization (living a fulfilled, purposeful life). Wisdom is the source of creativity, or human touch.

As a child, I learned how to read music but I was a horrible musician because I had no feel for music. My father had no formal training and lacked the knowledge to read music, yet he was a talented musician because he could play by ear. His inability to read symbols on a sheet of paper did not hinder his creative ability to combine sounds in a pleasant manner. I had musical knowledge, but my father had musical wisdom.

We increase our value over robots when we increase both our intelligence and wisdom. So, think in terms of not only improving your intelligence quotient (IQ) but also your wisdom quotient (WQ).

As I mentioned, wisdom is more nuanced than intelligence or logic. Once mathematics is understood, it is easy to determine the product or sum of two numbers. Historians, sociologists, and psychologists have studied 10,000 years of human history, and yet, predicting social outcomes is far from accurate. The best we can do is seek wisdom and try to avoid arrogance.

Action Plan

1. I can develop my _____ expertise into an expert decision support system.

2. I can overcome a major hurdle by purchasing a decision support system that does _____ .

3. I will use the following methods to improve my wisdom quotient:

   a. _____

   b. _____

   c. _____

Chapter 6

DEVELOPING YOUR HUMAN TOUCH

You are about to read one of the most important sections of the book, so it is appropriate to begin with a brief warning. To be blunt, the concept of human touch is somewhat, shall we say...esoteric, ambiguous, and amorphous. It is a downright enigma and therefore often ignored, because it is not easily measured. I encourage you to proceed with an open mind and to reread this section more than once. It is that important.

As you read this chapter, you might struggle with the concept of human touch being a viable defense against being made redundant by a robot. Most people want to hear a quick-fix answer, like "study science" or "become an engineer." Unfortunately, there is not an off-the-shelf, one-size-fits-all solution. Be wary of experts that claim they have an easy answer. None exist.

The problem is simple, yet the solution is complex. The threat is straightforward: The robots are coming for your job. Automation is faster, cheaper, and more precise than you will ever be. Machines are logical and become more productive with successive iterations of technology.

Human physical performance is finite and reaching a zenith. People cannot compete in the arena of increasing productivity. The solution seems elusive because the competitive skill needed is not in the physical realm.

Therein lies the key. The force needed to compete against automation is nebulous and unstructured. It is 180 degrees out of phase with the orderly robot. The solution is veiled in mystery, because the human comparative advantage is hard to quantify.

In the past, most of the value placed on human effort was machinelike in nature, even among skilled professionals. A worker was judged on his ability to perform a repetitive task. He was rewarded for being a reliable cog in the wheel. Even today, industrial age–conditioning makes it difficult to recognize the importance of human touch over mechanical output.

Unlike predictable machines, human actions are not logical. They act on emotion and can cause uncertain outcomes. Managers do not like uncertainty, and thus, their preference for machines over people. That is good news and bad news for the employee. As robotic applications advance, more jobs will be automated, especially those that are most repetitive. The loss of a job is bad news for the employee, but the good news is the jobs that will be available for humans will be less mundane.

It is paradoxical, but I believe the problem of emotional irrationality is the solution to combating the takeover by automation. The jobs that remain will be most appropriate for humans. They will value human touch over routine. That means there will be a demand for people that possess wisdom.

The confluence of emotion and logic is the formation of wisdom. Wisdom is superior to artificial intelligence. People can only have relevance in the age of automation by being more human, not more robotic. In short, you must learn to develop your human touch.

Follow Your Internal Compass

I am the father of six children. Although raised under similar conditions, from birth they each exhibited distinct characteristics and personality traits. Their interests and passions varied greatly. Some were studious, others athletic or artistic. My wife and I attempted to guide or nudge them in a particular direction, but each child's ultimate course was primarily directed by their own innate curiosity.

My youngest son Ezra is a case in point. Ezra has had an obsessive passion with cars all his life. As a baby, he ignored the stuffed animals, balls, and games that his siblings enjoyed. He only played with toys that had wheels. As a toddler, he wanted to sit in my lap and learn how to "drive" the car. Before he could identify colors or read, he knew all the major car brands. Ezra would point at a silver BMW and exclaim, "I like that red BMW!" The day I took him for his driving permit, he looked at me with all sincerity and said, "I've been waiting for this day all my life."

I tell this story to illustrate that unlike robots, humans can sense their purpose. They possess an evolutionary biological compass that points them in the direction of their capabilities. Developing human touch is a process whereby you align your outward actions with your internal sense of purpose. The difficulty is that there is not one correct answer or a single path. The way that you will know that you are on course is by following your biological compass.

Ezra was a boy with an innate love of cars; his biological compass was always pegged toward the goal of learning to drive. Likewise, your inner passion will lead you to pursue activities in your unique area of interest. Like a liquid, human desires flow to fill the volume that constrains their lives.

Automation will change our lives at a rapid rate never before experienced in human history. The robots are coming for your job. Employment situations will change in ways we cannot yet conceive. Your place in society will seem murky because your physical performance is no match for a robot. The curvature of employment has changed forever. You must flow to fill the new void. I believe the only path forward is to develop your instinctual sense of human touch.

What Is Human Touch?

What is human touch? It can be defined but it is harder to describe. Human touch is the ability to create. It is a singular human attribute. All living things can replicate themselves, but humans are the only species that can create beyond their own kind. From bicycles to hospitals to nuclear isotopes, mankind organizes matter to create things that did not previously exist. Robots are only as useful as the task they have been programed to execute. They can learn, but they cannot create. In its purest form, human touch is the creative process.

LEONARDO & LINCOLN

Mankind's creative prowess is infinite, from the silly to the sublime. It can consist of physical items like the Mona Lisa to conceptual thought as is written in the Gettysburg Address. Both of these examples show the extreme complexity of human creation. Leonardo da Vinci's painting expresses more than just a smiling woman; it is a work of art that has captivated human imagination for over 500 years. Likewise, the Gettysburg Address was not simply a eulogy, nor was it just a political speech. It is poetic rhetoric that has been used to bond and heal a chaotic nation for almost two centuries.

The Symbolism of Symbolism

The spirit of human touch is evident by our use of symbols. Symbols convey information in conceptual progression. Egyptologists claim to be able to read hieroglyphics. But do they really capture all the subtle embedded meaning? We have no way of knowing, because the cultural context is extinct. Context is everything. My grandfather used the word "gay" to describe happiness; I interpret it as a sexual preference.

The word "hieroglyphics" comes from the Greek "hieroglyphikos," which literally means "sacred carving." The word "hierarchy" shares the same entomology; it means the "rank in sacred order." Today we use the term to describe a system of organization. Symbols convey information in a hierarchy of cognition. Recognition of symbolic meaning is not based on mathematical logic (1+1=2), but rather emotional interpretation.

What do you *feel* when you see the logo of your favorite sports team versus the logo of an opposing team? Does a diamond ring simply represent a piece of jewelry? Is your job a means of earning compensation or a purposeful calling?

Consider a three-pointed star within a circle, the trademark logo of Mercedes-Benz. The emblem represents much more than simply a brand of automobile. Depending on your unique cognitive context, it could represent luxury or elitism.

Robots can be programmed to interpret symbols, but they cannot *feel* the meaning. You can. More importantly, you can create an impression for others to feel.

Feelings

Some will argue that robots will become sophisticated enough to develop interpersonal relationships with people. Artificial intelligence may endow robots with emotional bonding capabilities, but I do not believe that humans will reciprocate. Observe how people have bonded with technology in the past. Technology of old is not impressive to you but it is always life changing to the people who first adopted it. Indoor plumbing, electricity, refrigeration, television...the list is endless. All these technologies were revolutionary and greeted with much enthusiasm.

Despite peoples' love and connection with a new technology, it is fleeting. Just like a one-night stand, the thrill is gone when a new, sexier player emerges. In the 1950s, consumers loved the tiny little black and white screen on their television. When color was introduced, they could not wait to upgrade, just as they did when cathode ray tubes were replaced with flat screens. This is the course of technology—the new rapidly replaces the old.

When a new smartphone is introduced, consumers camp out overnight to be first in line to purchase. They do not mourn the loss of the old phone. People cry when their pet dies, but they joyfully throw out old appliances to upgrade for something better. This is the state of the human condition.

People fall in love with living things, not inanimate objects. Aside from those with social pathologies, that is the nature of humanity and it will not change. The robots are coming for your job, but they cannot replace you if you are loved. People that learn to use their human touch to create will prosper; those that create emotional bonds and feelings with others will thrive.

BLACK & WHITE TV

Television is an easy way to gain insight into human nature. Watch some old black and white television dramas from the 1950s. Get some popcorn, your favorite beverage, and enjoy some old shows like *Perry Mason* or *Dragnet*. Note how drastically different their lives were without mobile phones, air conditioning, or the Internet. More importantly, notice how human nature has not changed. The dramatic theme revolves around something like murder, adultery, or greed. It has always been that way, and it always will be. Instead of fighting human nature, learn to identify and profit from it.

Wisdom

As we've explored previously, wisdom is the result of knowledge guided by human experience. It is the product of logic and emotion. Our creative human touch has its origin in our wisdom. The divide that separates human wisdom from robotic artificial intelligence is vast. Think of it as a demilitarization zone that will provide defensive cover against an impending robotic invasion.

Robots think logically, as characterized by Hollywood stereotypical humanoids decrying, "does not compute," or *Star Trek's* Mr. Spock saying "highly illogical." Counterintuitively, the robot's strength of precision and logical efficiency are its Achilles' heel. What the robot cannot compute, you can feel. That single advantage will set humans apart from the onslaught of automation's superior performance, much as the simple anatomy of opposable thumbs sets primates apart from much stronger mammals.

Automation is highly productive at routine tasks, but it cannot create beyond the limits of its inherent programing. Advances in artificial intelligence will enable robots to learn, but not create. It is a subtle difference, but the defining factor of human value.

Data, Things, People

The Bureau of Labor Statistics uses the Standard Occupational Classification to categorize over 800 distinct job types. It is not practical to argue the effect of automation on each of these jobs. The short answer to the question of what jobs will be most impacted by automation is those that are most routine with the highest pay. This concept can be easily characterized by the highly compensated medical professional that performs a routine task, like administering anesthesia or reading an x-ray. These are complex tasks, but they are also predictable enough to eventually be fully automated by a computer.

How can you determine if your job is at risk? Rather than considering over 800 occupations individually, I think it is more constructive to broadly look at the three critical tasks that involve work. In one way or another, we all work with data, things, or people. The more complex the nature of the work, the more these three tasks become intertwined. A dishwasher would likely only deal with things (dirty dishes), while a patent attorney would likely delve deeply into all three tasks.

"Data" primarily involves working with numbers. Accountants, actuaries, analysts, and computer scientists all deal with data. Since data is mathematical in nature, it can easily be automated with an algorithm. Jobs that involve low-level calculations will be easily replaced. An accountant that prepares tax returns might be an example of a job function that will be made redundant. On the other hand, those that use data to create content will be in high demand. Obviously, this would be the computer scientist that writes code, but it will also include those that can interpret raw data into meaningful results; for example, a marketer that interprets consumer-purchasing data and translates that into sales strategies.

"Things" refers to working with inanimate objects. Carpenters, pilots, and electricians all work with things. The effect on jobs in this category will be less pronounced than in the data category. Replacing an electrician who wires new construction could occur if homes and buildings were prefabricated using factory-built modular construction. However, it would be very difficult to automate rewiring an old house. Piloting an aircraft can be automated, as the use of pilotless drones has shown. However, due to concern for safety and public perception, it is unlikely that commercial aircraft will become fully automated anytime soon. What is more likely is that flight crew headcount will be reduced. Just as onboard radio operators, flight engineers, and navigators have been replaced with technology, eventually copilots will be redundant. People who work with things will find opportunities by integrating many items into one useful device or service. This includes the electrician who can wire your house with easy-to-use security and entertainment systems, or the cruise ship captain who not only pilots the vessel but also creates a memorable experience by interacting with the passengers.

"People" jobs obviously refer to tasks that primarily involve interaction with humans. Low-level interaction jobs, like bank tellers or checkout cashiers, will be easiest to replace. Highly interpersonal services, like addiction counselors or trial lawyers, will be harder to automate. Those that can nurture strong emotional ties with their customer base will be least likely to be replaced by a robot.

Ultimately, the most secure jobs and highest incomes will be earned by those who can integrate data, things, and people together. Think Mark Zuckerberg and Facebook. That is the manifestation of human touch, aligning standard tasks (data, things, and people) with unique human characteristics (leading and supporting traits) and creating something new (product or service). As we close out this section on thinking like a human, take time to reflect on where your biological compass is pointing.

Action Plan

1. Since childhood I have been fascinated with:

_____

_____

_____

2. I daydream about:

_____

_____

_____

3. Time flies when I perform the following activities:

_____

_____

_____

4. My favorite hobbies are:

_____

_____

_____

5. Activities I enjoy at work:

_____

_____

_____

6. List the activities that appear multiple times above:

_____

_____

_____

Your internal compass is likely pointing toward the activities listed in #6. Your current level of happiness is probably proportional to the extent that items in #6 match the activities you frequently perform at work. If you feel unhappy, you should seek employment in areas that allow regular participation in activities listed there.

PART TWO
Entrepreneurship

Chapter 7

THINK LIKE AN ENTREPRENEUR

I stress the concept of following your inner compass because it is easier to work with nature than to work against it. You are on the right track when you feel happy. Like the lumberjack trying to split a log, happiness occurs when your outward actions cut in the direction of your inner grain. That does not mean you will not experience pain or discomfort. If what you are trying is not working, try something else. Push yourself. Pursuing happiness does not mean you will never feel stress. The trick is to make sure you are fighting the proper battle. Do not try to push a string; it's futile and a waste of energy. Push against the complacency that is restricting your growth. You will likely find that to create newness, you will need to get out of your comfort zone and experience new things.

Think like a human was probably an easy concept for you to grasp. The logical conclusion is that we can never be as productive at repetitive tasks as a robot. So to achieve, we must focus on our unique strengths—our humanity.

The next cognitive instruction may make you feel more uncomfortable. *Think like an entrepreneur, not an employee.* An entrepreneur is someone that organizes a business and takes on greater than normal risk. Even if you never plan to become self-employed, you should start thinking like an owner, because ultimately, you are the owner of your own career.

There are comparatively few entrepreneurs, because most people do not want to take on the burden of risk, fearing failure. Employees prefer the security of a regular paycheck. If you are paralyzed by this primal fear, you have to first acknowledge it and then overcome it by trying new things.

You are never too old to try new things. I can provide many examples of famous people that did not become successful until turning to entrepreneurship late in life. I personally was a late-blooming entrepreneur. But the example that I want to use for trying new things involves even more of a base emotion: a love-hate relationship. All my life, I hated animals. Not that I was ever cruel to them, I just wanted them far from my everyday life. I thought owning a pet was ridiculous. My favorite animal was a cow, when it was served as a steak.

Fast forward to a few years ago. My kids were growing up and leaving the nest. I needed a distraction, someone to play with. Crazy idea—I got my first puppy when I was in my early fifties. He has become my best friend. I am a certified dog lover. I still hate cats, but that works out well, because so does my dog.

I DON'T WANT TO BE AN ENTREPRENEUR

The automation age will create extraordinary opportunities for those who want to become self-employed. But maybe that is not your desire. No problem! To prosper in the future and beat out the robot, you will not have to become an entrepreneur. But you will need to think like one. Start now to challenge yourself to think creatively. Think in terms of adding value, solving problems, and creating new products and services. Your employer will be able to purchase a robot that can quickly accomplish routine tasks. Creative people will always be in demand.

Start Narrow, Start Now

How to do you start your enterprise? Should you specialize or be a generalist? Should you sell locally or globally? Should you start with a service and follow with a product, or vice versa? Who knows? Just start somewhere. If it does not work, try again.

I knew a very successful dermatologist with an extremely lucrative practice. He worked very relaxed 9-to-5 hours and was never stressed. When his fellow physicians were on call or working long shifts in the operating room, he was enjoying time with his family. I asked him about the secret to his success and how he had structured such a balanced lifestyle.

He told me that in medical school, he had many prestigious options and qualified for just about any specialty of his choice. He chose dermatology because it seemed like the perfect sustainable business model—he could not cure anyone, and he could not kill anyone.

Superficial? Perhaps, but he was highly compensated for doing what he loved, and his patients were exceptionally loyal. He never cured nor killed them, and they happily kept coming back. He made a fortune.

Will your concept be successful? Will it generate millions or billions in revenue? You will never know until you launch your idea. Start small, scale up, and make corrections along the way. That simple path will lead to your eventual success.

Personal or Transactional

Think about your desired income and the profit needed to attain that amount. How will you go after that market share? You have three options:

- Receive one paycheck from a single employer.

- Conduct a small amount of transactions for a high-priced item or service.

- Conduct many large transactions for a small-priced item or service.

Your personal preference and the market response will most likely dictate what direction you take. All three methods work. I have a friend who earns a nice income making a small margin on lots of transactions. Her business model is automated to be transactional, not personal. Customers pay with a credit card on her website and she does not have to offer individual service. It is the nature of her business model.

I established my firm on a personal concierge model where I have a small number of clients but the fee is several thousand dollars. I know each client personally, and if they have a question, they call my mobile phone day or night.

My friend has "customers." I have "clients." It is a personal preference. We both make nice incomes and have satisfied patrons.

You can earn $200,000 two ways:

- Two thousand $100 transactions

- One hundred $2,000 transactions

Which do you prefer? Which will your market allow?

Employ Robots

One reason I am so optimistic about the future is that automation will minimize monopolies, and the barriers of entry to aspiring entrepreneurs will be small. Competition will abound. That is good for the consumer, and that is good for the achiever.

As outlined in a previous exercise, you should have some thoughts about your weakness or disadvantages, and what types of technology you can "hire" to minimize them. Automation will make robots plentiful and affordable. Use them to your advantage.

Think broadly about automation being technology. Your robot may be something as simple as software that prepares your taxes or as complex as a diagnostic expert system. Use technology appropriately and sparingly.

New and improved technology will be developing at an astonishing rate. Do not buy more than what you currently need because it will soon be obsolete. Settle for nothing more than a three-year payback, or preferably 18 months.

Get started now with the minimum investment. My business is highly profitable and one of the reasons is that I do not waste money on unnecessary infrastructure. When my firm needs more computer storage memory, I buy more. As traffic increases on the website, I upgrade to greater bandwidth. I do not need an elaborate, expensive super computer to trade stocks. The most expense piece of equipment in my office is the Herman Miller Aeron chair that I sit in. It costs about the same as a high-end smartphone. The ergonomic design is over 20 years old and is still the best on the market. My chair will last a lifetime as long as it is not abused. An iPhone is obsolete in about 18 months. The chair is a more valuable investment.

In an age when high-tech financial algorithms and "robo" financial advisors are being hyped, you might think that a money manager like myself would need a flashy website to be competitive. Not necessarily. Actually, my firm's website is rather unimpressive...by design.

I started my firm on the premise that it would grow organically through referrals from satisfied clients. I was not interested in mass marketing to the general public. I did not want a colorful website because it would not be authentic to me. I am not a flamboyant guy. I drive a Subaru, not a BMW—by choice, not by necessity. My wealth was acquired though successful stock market investing and frugality. That is the financial gospel I preach. The clientele I serve is the middle class self-made millionaire.

Initially, I set up a rudimentary website powered by WordPress and eventually planned to make it look more impressive. But, serendipitously, the unimpressive design proved a powerful marketing tool, or more descriptively, a prospect qualifier. The simple design of the website acts as a self-selecting filter to differentiate potential clients.

My target market, the rugged, self-made person, is not deterred by the bland design. They are impressed with the content of my stock market commentary and, thus, they seek out my services. On the other hand, ostentatious people do not call me. It saves me the trouble of having to sort through prospects that would not be a good fit for my firm. It acts as a robot receptionist that spares me from wasting time with inappropriate prospects. The "design" is free, but to hire a human employee to perform the same gatekeeper function would be at least $30,000 per year. Priceless.

Automate wherever you can, but only when the financial payback can be captured in a relatively short period of time. What cannot be done with technology, farm out to capable freelancers. Minimizing the number of employees will not only keep overhead low, it will greatly reduce complexity.

Funding

Financial markets move in cycles. For over 30 years our economy has been experiencing a long-term trend of declining interest rates. Presently, short-term government debt is yielding near zero or, in some cases, negative interest rates. Whether rates will stabilize or move lower is uncertain, but I do not believe they will rapidly escalate back to historic highs any time soon.

That means that money is "cheap," and likely to remain so. That is good news to the aspiring entrepreneur who wants to fund her growing enterprise. More good news is that sources for funding keep increasing. Credit card debt was once the only loan available to many budding entrepreneurs. Today, there are numerous sources for startup capital, including nontraditional crowd sourcing and peer-to-peer lending.

There is one secret to receiving a loan, regardless of whether it is from a family member, private equity investor, or crowd sourcing: The creditor must believe in your ability to deliver.

Just like a resume or curriculum vitae must demonstrate to a prospective employer that you have creative talent, your loan application must demonstrate a business that can produce profits. How is that done? By demonstrating demand for your product or service. And that is accomplished by marketing some tangible form of your product.

Proof of Concept

So, it is a bit of a chicken-and-egg conundrum, but the bottom line is that to receive funding to grow your enterprise, at minimum, you need enough seed money to develop and market a functioning proof of concept (POC). The POC does not have to have all the final bells and whistles. A workable prototype is usually sufficient, as long as it displays the potential value of a fully functioning product. Think in terms of the software industry's practice of releasing a partial program in a beta version. The innovator never gets it right in the first attempt.

It is an iterative process, just like Thomas Edison's one thousand attempts at inventing the light bulb. Consider the success of Apple—they did not invent the first of anything—computer, tablet, MP3 player, smartphone... nothing. What they did do was take an existing product and make it increasingly better.

A fast and efficient way to get started is to develop a POC. From table-top 3-D printers to outsourcing manufacturing in China, technology has made it easier than at any time in human history.

Begin by turning your idea into a tangible item. If you are an artist, that might be creating a painting or a song. If you are an electrical engineer, that might be creating a light saber. Just get off your backside and move your concept from an idea to a thing.

Then, sample it. Give it away to family, friends, and strangers. Willingly receive feedback. Improve and innovate based on that real-world user feedback. Do it more than once. The cost to improve a POC is much less expensive than making modifications to a commercialized product.

The next step is to take the POC to market and see if anyone buys it. Again, this does not have to be the final product with all the bells and whistles. At a minimum, it should be a viable, functioning mock-up of your final design, and it should be completely safe to operate. Get it out there and see if it sells.

There have never been more opportunities, such as the following, to establish an initial market for your product or service:

- Consignment sales at a local specialty retail store

- Preorders via a crowdsourcing site

- Free advertisements in Craigslist or small-budget ads on Facebook

- eBay, Etsy, or Amazon storefront

- Product placement on a specialty blog, podcast, or YouTube channel

I will not belabor the methods of getting your product to the market. Those sources are as wide and varied as your unique product will be. The point is that there are now more options available than ever before, and they are rapidly expanding. To be blunt, if you cannot determine how to initially get your product in front of early consumers, then you are probably not creative enough to develop a viable product. Do not waste your seed money hiring a marketing expert.

To some extent, it really is like building a better mousetrap, or the quote from *Field of Dreams*: "If you build it, they will come." It will take a little nudging and probably some incentives, but if you have a good POC you will find early-adopting consumers. Initial sales lead to increasing sales, which result in positive cash flow and, ultimately, sustainable profits. It can really be that simple, but it all starts with the first sale.

Future Tax

It may seem premature to raise the spectrum of tax issues before you have even launched your concept, but I do not think so. In the spirit of Stephen Covey's "begin with the end in mind," I think you should include tax strategies from the very start, if for no other reason than startup costs themselves are tax deductible, even if the project fails.

Looking into the future, I feel confident about making only two firm predictions: 1) the robots are coming for your job; and 2) the government is coming for your money.

I do not say that as some reactionary, crazed tax protester. The fact of the matter is that both of those outcomes are intricately connected. The first, robots creating unemployment, makes the second indisputable. Higher unemployment will require more government redistribution of wealth from the employed haves to the unemployed have-nots.

If you plan to earn income in the future, you need to consider tax strategies now. Specific recommendations are beyond the scope of this book. I mention the subject only to introduce you to the need. It is generally the farthest thing from the entrepreneur's mind.

Tax consequences are not only important from a fiscal standpoint, but also from a lifestyle perspective. This is yet again an advantage the employer has over the employee, and often overlooked.

Employees have taxes deducted directly from their wages before they even receive their paycheck. The options for legitimate employee tax deductions are sparse. The typical employee spends about one third of his life at work and has the fruits of his labor taxed near 50 percent after all the jurisdictions take their share: federal, state, local, income, payroll, sales, property, capital gains, gasoline, tolls, usage, excise, etc. All this will get worse as income inequality widens.

The employer spends just as much time at work as the employee; however, as an owner, he can qualify for many lawful tax deductions as a result of legitimate business expenses. As an employee, you spend your free time pursuing recreational activities, hobbies, and vacationing at your own expense. The owner is never clocked out; she is the business. Her time out of the office is spent entertaining clients and developing personal skills at the company's expense. When the employee goes to Hawaii, it is a self-funded vacation; when the owner goes, it is a business trip to visit a client.

The difference between pre-tax business expenses and after-tax employee income can be substantial. To fund a flight and hotel for a Hawaii vacation that costs $2,000, an employee with a 30 percent tax rate would have to earn $2,857 ($857 goes to taxes before the employee receives their $2,000). A $2,000 business trip to Hawaii for the purpose of visiting customers or attending a professional conference would be expensed pretax at $2,000. (Consult your CPA; this is given as an example and not intended as tax advice.)

As a budding entrepreneur or free agent, you should consider ways to construct your enterprise so that it complements your desired lifestyle. Just like tweaking your early POC designs, the foundation of a lifestyle business is easier to establish at the beginning than at the end.

STRATEGIC RELOCATION OR HUNTING CAMP?

Sam Walton, the founder of Walmart, was from a small town, but he was no hick. He grew up in a small college town where he also attended university and graduated with a degree in economics. During World War II, he served as an Army intelligence officer. His father was a banker and his father-in-law was a prosperous rancher.

As an adult, Walton had lived and worked in several mid-size cities, like Tulsa, Des Moines, and Salt Lake City. The question is, when he decided to go into business for himself, why did he locate to obscure rural Arkansas? Was it because real estate was less expensive? Was it part of a niche strategy to meet the retail demand of underserved small towns? The likely reason is that it was close to his wife's family and, more important to him, western Arkansas is renowned for quail hunting.

Walton's passion was bird hunting. His hunting dogs were iconic. Walmart's branded dog food "Ol' Roy" was named in honor of Walton's favorite dog. When asked why he drove a pickup truck, Walton replied, "What am I supposed to haul my dogs around in, a Rolls-Royce?"

Walton was famous for visiting Walmart stores in out-of-the-way locations that just happened to be near excellent quail hunting. He usually managed to work a hunting trip along with a store visit. Or was it the other way around? Either way, it did not matter. Walton loved his stores, his employees, his dogs, and his birds. He led a balanced life that incorporated them all. We would be wise to follow his example.

Establishing a lifestyle business is also critical to keeping balance in your life as you pursue the ultimate goal of happiness. Regardless of tax incentives, if you are going to spend at least a third of your life at work, you should be doing what you enjoy. For example, perhaps you are obsessed with basketball but you stand only five feet tall. It is unlikely that you will be drafted by the Los Angeles Lakers as a point guard. But you can still incorporate your love for the game into a viable career:

- Sports commentator

- Cameraman

- Coach

- Recruiter or talent scout

At the very least, you could start any type of successful business venture and entertain your clients at basketball games. I have known many lackluster salesmen that crafted a lucrative corporate career around their ability to play golf with customers.

A final note about monetizing your human touch by thinking like an entrepreneur, not an employee: If you can successfully implement the above methods into your plans, you will inevitably build a resilient enterprise. The benefit is that even if the robots do not come for your job, like asbestos, you will be fireproof.

My firm currently employs just one individual. He is brilliant and irreplaceable. Even if he were not, it is unlikely I would ever fire him. My firm is a solo practice: the only employee is me.

Action Plan

1. My business idea (or idea to highlight your competency at work) is based on the following concept (product or service):

   _____

   _____

2. Do I have the means (money & resources) to develop a simple prototype that captures the essence of my concept?

   a. YES: Proceed to #3.

   b. NO: Start over at #1. If a simple prototype cannot be created to highlight the genius of your concept, the odds are slim that the idea can be developed.

3. The prototype will:

   a. Contain these key features that demonstrate the value of my concept:

   _____

   _____

   _____

b. Not contain these features, which will be included in the final design:

_____

_____

_____

4. The cost to develop my prototype is: _____

5. Is this cost justifiable (am I willing to put my hard-earned money into this idea)?

   a. YES: Proceed to #6.

   b. NO: Start over at #1.

6. Develop the prototype so that it can be sampled or displayed to likely users.

7. The feedback is:

   a. Positive: Proceed to #8.

   b. Indeterminate; with some tweaks, the prototype could be improved: Start over at #2.

   c. Negative: Start over at #1.

8. My concept seems viable. I will proceed in a similar manner to have a small quantity of working prototypes manufactured and attempt to sell them.

Chapter 8

THE CREATIVE ADVANTAGE

There are many theories about the source of creativity. I believe that creativity results when right brain, the feeling hemisphere, is tempered by the practical left brain. This is obviously a simplified generalization, but it serves as a model to illustrate the point.

The left hemisphere of the brain processes logical functions, like mathematics and language. These logical functions are things that a computer algorithm can easily mimic, either now or at some point in the future. They include voice recognition, language translation, innumerable mathematical calculations, and any functions that result from rule-based logic (if a=b and b=c, then a=c).

Siri can answer questions that have a logical answer, but not complex emotion-based queries. She can tell you where the closest Starbucks is, but not whether you are in the mood for a hot or cold latte. Likewise, expert decision support systems will be extremely efficient at capturing human knowledge and answering rule-based questions:

- Medical: Is the tumor malignant or benign?

- Legal: Has a precedent been established in prior judicial decisions?

- Mechanical: Can a beam withstand the applied load?

These functions are all rule based, and thus predictably routine. The question may be complex, but the solution can be reduced to a mathematical algorithm.

Conversely, creativity occurs outside the limits of a rule-based system. For example, Einstein disregarded the rules of Newtonian physics to develop the theory of relativity. To envision relativity, Einstein constructed a thought experiment by imagining what it would be like to ride on a beam of light.

This is where the nuance of the right-brain hemisphere comes into play. Exactness is the domain of the left brain, like the calculation of a physics equation to determine a precise robotic movement. On the other hand, the right hemisphere of the brain interprets visual and auditory clues, transforming them into feelings. From these feelings, unique human emotions emerge that create works of imagery, like art or music comprehension. Creative imagery was the essence of Einstein's genius, something a robot does not possess.

Likewise, a robot can use facial recognition software to identify the biometric features of an individual. However, the human brain not only identifies the features; it also has an emotional response to the perception of beauty of the individual. For example, when you look at someone's face, the left-brain function will calculate the geometric proportions of that person's features, and the right brain will determine if you find those physical attributes attractive. Beauty is not in the eye of the beholder; it is in the right hemisphere of the brain. Computers can be programed to recognize facial features and attribute those features to accepted norms of beauty. However, computers cannot interpret the emotional *passion* of that beauty.

The difference between logic and passion occupies the same untouchable humanity as wisdom. Think of what a robot can create with a 3-D printer. Now consider what the genius of Leonardo da Vinci could have done with it. There is no comparison. Leonardo would never have been made redundant.

Constraints of Logic

Einstein did not allow his sense of creativity to be restrained by conventional thought. Likewise, the rap artist does not follow the rules of a classical pianist. Robots can easily play or write musical pieces once the style has been established, just like robo-journalists can follow a media template or robo-advisors can execute a stock trade. It is simply a matter of executing a rule-based algorithm. It is nothing more than following a trail that has already been blazed by a prior human. No creativity needed, just mimicry. Human touch is superior because it can create original content to pioneer a new style.

FINDING BALANCE

Elvis Presley was the king of rock and Michael Jackson was the king of pop. These music legends lived epic lifestyles. They created musical genres that defined the age in which they lived. Yet, despite their fortune and fame, these men led troubled lives often characterized by addiction and pain. This is frequently the case with creative genius. An internal battle between good and evil; a struggle to balance pleasure with pain. Keep this in mind as you seek to develop your own creativity. To break through conventional thinking, you must challenge the status quo and cultivate right-brain thinking. However, Aristotelian moderation is required to keep the extremes of creativity in check. Wisdom results when emotion is sufficiently bridled with reason.

Creativity is a compromise between the two hemispheres of the brain, one based on logic and the other based on emotion. The mixing of the two results in our human ability, thus creating an infinite combination of unique characteristics. An architect is more "right brained" than an engineer; an artist is more "right brained" than an architect. Like a good partnership, the relationship between the two brain hemispheres is not always equal but must work in harmony. Too much of one thing is seldom good, especially when it comes to emotion. The logic of the left brain serves as a taskmaster to keep the creative right brain in check.

I like to think about creativity as a struggle between the two brain hemispheres: logic versus feeling. In a sense, it's the age-old battle between good and evil, where moderation is usually the best alternative. Think of the stereotypical absented-minded scientist or the eccentric artist. We expect genius to be quirky. That disregard for the conventional makes the genius excel above her peers. It is the essence of creativity and it is what robots lack. As stated earlier, artificial intelligence is not artificial wisdom.

Was Picasso a genius or did he paint gibberish? Was Dali's work profound or pornographic? A robot cannot tell, does not care, nor can it create the originating style. Art, like all forms of creativity, resides in the domain of the human mind. It is hard to describe, and it cannot be logically quantified, nor easily reduced to an algorithm. Genius is purely human. You will maintain the upper hand over automation by creating original content. But caution is advised—we often hear of the starving artist but seldom the starving engineer. The crucial element is to find harmony among your talents.

Original Content

Content comes in many forms: works of art, scientific discoveries, and simple acts of kindness. Original content may be enhanced by automation, but not replaced. Consider the skilled surgeon, whose expertise is rapidly becoming commoditized by robotics like Intuitive Surgical's da Vinci Surgical System.

The system provides the surgeon with a magnified 3-D, high-definition internal view of the patient's body and precision control of tiny instruments mounted on flexible robotic "wrists." Unlike traditional laparoscopy, where the surgeon directly manipulates fiber-optic instruments, the da Vinci Surgical System translates the surgeon's hand motions into smaller, more precise movements. Human error can be minimized by correcting jerky movements and limiting intrusion from outside of the surgical boundary.

The da Vinci Surgical System can make a good surgeon capable of delivering great results. It is a win-win for both patient and physician. The patient receives extraordinary care from a surgeon of average skill, thus the supply (of qualified surgeons) is readily available to meet demand. Likewise, it is a win for physicians, both the good and the great. The run-of-the-mill general surgeon can step up his game from performing simple appendectomies to complex neurosurgery. The gifted surgeon can spend his time on harder cases and creating advanced procedures. In fact, much of his time will no doubt be spent developing robotic systems that mimic his superior ability, which will enable less skilled doctors. Synergy occurs because physicians are able to increase their level of productivity. Everybody wins. Well, almost everybody.

The loser is the surgeon who remains stagnant by maintaining the status quo and not adopting new technologies. The loser is also the surgeon who ignores customer service. Remember, "content" comes in many forms, including simple acts of kindness. Superior customer service can be branded as content when it is tailored to the customer's unique needs.

In the past, a renowned heart surgeon could afford to have poor bedside manners because his skill was in such high demand. When technology commoditizes complex medical procedures, physicians with prima donna attitudes will be avoided. The technologically enhanced doctor of average skill with a warm, caring demeanor will be highly sought after. This example is universal in application; it applies equally to cardiologists or carpenters. When given a choice, people place a high value on kindness.

Building Personal Relationships

The act of creating is not limited to physical products and services. It extends to intangibles like kindness and love. As stated previously, the concept of human touch can be murky. This is one of those fuzzy areas. Love is an emotion, and it cannot be quantified with exactness. There is no discovery mechanism to determine its monetary price. Nonetheless, love is highly valued. Those who can create emotional bonds like kindness and love will be highly regarded and rewarded. The robots are not coming for those jobs.

BE A ROCK STAR

My mother was born in the midst of the Great Depression, so she had a different perspective compared to you or me. Whenever she made a bank transaction, she always went inside to be served by a bank teller. I was a teenager when ATMs were launched and preferred their convenient 24-hour access. As such, I was an early adopter of online banking.

My mother loved talking with bank tellers because, for her, it was a pleasant social interaction. To me, it is an avoidable annoyance. It is not that I prefer to deal with a machine over a human, I just prefer to deal with an efficient machine over a time-wasting human.

If the bank teller was Warren Buffett or Heidi Klum (think George Clooney for the ladies), I would most likely stand in a long line to talk with them. But when I go into a bank, I never get Warren or Heidi; instead, I get Nancy-Know-Nothing.

The bank could grow its business and provide a better customer experience if human employees were enabled with big data expert systems. A computer algorithm could prompt a bank teller to provide customers with meaningful information based on the customer's spending and saving habits. Not simply upselling bank services, but actually providing customers with meaningful recommendations. "Mr. Pugliano, if you switched to free overdraft protection, you could carry a smaller balance in your checking account and earn $200 in additional interest."

An ATM or online banking service might offer convenience, but it does not build customer loyalty. A *perceptive* human bank teller who provides valuable money or time-saving tips would help to establish the bank's reputation as a trusted resource. Banks that build personal relationships are more likely to retain customers and sell those customers a wider range of financial services.

Regardless of what type of work you do, strive to provide warm, caring, and informative service to those you come in contact with. If you want to be preferred over a robot, be a *rock star*, not a Nancy-Know-Nothing.

Creating Value

The robots are coming for your job if it is routine or if it can be reduced to an algorithm. Automation will replace the simple and the complex. No profession will be spared redundancy. Automation will not replace the unknown, because it cannot. The unknown is not routine. The unknown cannot be formulated into an equation because, well, because the outcome is...unknown.

What is unknown? The new. Things that are created afresh and never existed before. The telegraph was once new, as was air conditioning. What will be new tomorrow? I have no idea. It is unknown.

What I do know is that professions and careers that are on the frontier of developing valuable new products and services will not be replaced with automation. Therefore, the careers and professions that deal with creating new value will always be in demand.

The good news is that everyone can participate, if they create value. There will be need for carpenters and cardiologists, as long as they are creative. As long as the craftsman, physician, attorney, Sunday school teacher, etc., is producing a valuable product or service, they will *create* demand.

New products and services that we cannot even comprehend will be available in the near future. They will be created by entrepreneurial pioneers, not status quo bureaucrats. New value will be created rapidly, creating opportunity for even newer things. Technology and robots will make the creative even more productive, thus forming an upward spiral of infinite opportunity...for the creative.

Da Vinci was creative. He conceived of things that were ahead of his time, like submarines, flying machines, and the study of human physiology. Imagine if he had a 3-D printer or a DNA sequencer. People of his intellect will be born in the future, and they will have access to these amazing technological tools. You may not be a da Vinci, but that is okay. The innovative genius of others will improve your life and provide you with opportunities to advance in your area of interest—provided you create something at your level. I did not create the products and services of Microsoft, Apple, or Google, but I used their technologies to advance my career and improve my life.

Critical Success Factors of Creativity

While I cannot tell you specifically what you should create, I can propose two enabling factors to assist in your discovery process. The first critical success factor of creativity is understanding value, specifically that the perception of value lies with the consumer, not the creator. This might seem obvious, but it is extremely complex because personal preference is extremely subjective. Like fashion, it can also be fleeting. What you perceive as a valuable service might be worthless to me.

The National Football League (NFL) has revenue in excess of $13 billion; its commissioner never plays in a game, yet receives compensation of around $30 million. Sports enthusiasts have fanatical attachments to their favorite team. Yet, the NFL is valueless to me. I hate football. I have no preference for it at any level, professional, college, or high school. You probably love it.

The context of value must be from the consumer's standpoint, not yours. This is extremely difficult for many inventors and creative people to comprehend because they have such a deep personal attachment to their own work. The artist loves her creation because it originated within her. She feels the intrinsic connection because it is literally a piece of her mind. Yet the painting goes unsold because others did not perceive the value. Like the struggle between pleasure and pain, the creator must find a balance between her own interest and what others value.

Ironically, the second critical creative factor is authenticity. While you must create for the benefit of others, your work will not be extraordinary if it is not authentic to you. If your creation is simply a replication of another's work, then it is likely a commodity that can be mass-produced by a robot.

The advantage of authenticity is that it is not easily replicated. If you create unique value, competitors will find it difficult to gain an edge. Consider the talented educator. He uses the same factual information as any other teacher. Yet, his students learn more because he presents the material in a fashion that inspires them to learn. His presentation methods are unique to his personal style of communicating. The more distinct his mannerisms, the harder for an impersonator to mimic them.

Comedian Ellen DeGeneres can tell a funny story that captivates your attention and has you laughing with tears. I could tell the same tale and it would fall flat. The content of her story might not be that funny, but her presentation style is.

Happiness and the Human Touch

Human touch is the ability to create. This is how people find meaning in their lives. Educators create scholarly students. Comedians create jokes. Parents create joyful homes. We all can create in our sphere of influence, in our area of interest. Opportunity abounds. The only constraints are our own talents.

Is your creativity on the right track? Listen to your inner voice; observe what direction your biological compass is pointing. It is your self-correcting feedback mechanism. If your work leaves you feeling unsatisfied and stressed, you are most likely not pursuing a successful path. Even if your output is world-class, you will eventually burnout if you are unhappy.

Happiness parallels human touch. You are likely to be most happy when you are engaged in projects that are aligned with your talents and interests. Your biological compass will always point toward what makes you happy.

Pursuing happiness will energize your human touch. The result will be the creation of authentic products and services that are highly valued by others. As the creator, you will be monetarily and personally rewarded. Your happiness will increase, which will further improve your creativity. You will have created a positive feedback loop that is more productive than a supercomputing algorithm.

Action Plan

1. I am happy when I perform the following activities:

    a. _____

    b. _____

    c. _____

2. The above activities can be used to create:

a. Services:

1. _____

2. _____

3. _____

b. Products:

1. _____

2. _____

3. _____

3. I can create personal connections with people by:

a. _____

b. _____

c. _____

4. How can I become a kinder person?

a. _____

b. _____

c. _____

Chapter 9

AN ECONOMIC PRIMER

Technologies will come and go, but economic laws will remain in force because they are based on human interaction. When two people engage in exchange, it forms a market. The medium of exchange is based on the technology of the day, like the barter system of the Paleolithic days or Federal Reserve currency of today. While the medium changes, the dynamics of commerce remain. Supply and demand always determine price, whether the currency is wampum or Bitcoin. Markets always exist, even in societies like North Korea that try to eliminate them. Markets exist because they are an expression of our human nature.

Four Factors of Production

Economists separate production into four categories: capital, land, labor, and entrepreneurship. For our purposes, the trilogy of capital, land, and labor can be thought of as tools that the entrepreneur uses to create a profit. Capital, land, and labor are inanimate commodities that are organized into valuable products and services by the entrepreneur. Entrepreneurship is essentially what I describe as the creative ability of human touch.

A classical economist would argue that labor is not an inanimate commodity because labor is the input of human effort. Therein lies the paradigm shift of the coming automation era. It is my contention that "labor" will not exist in the economy of the future, because it never really existed in the past. The concept of labor is a misunderstanding…but more about that later.

Capital

Most people think of capital as money. But in an economic sense, capital is the tools used to transform natural resources into products and services. So, money (profits from the operation of a business) is reinvested back into the business to buy tools that will help the business more efficiently produce new products and services.

- Cavemen used a stone-tipped spear to improve their ability to hunt.

- Machinists used a lathe to shape a metal rod.

The old rules of capital allocation will apply to the automation era. Robots are simply tools that entrepreneurs will use to transform natural resources into new products and services. Directional drilling has made shale oil reserves economically attainable; deep sea robots will make mining the ocean floors practical.

Land

In an economic sense, land is a euphemism for natural resources. Land is the physical real estate, as well as the resources contained therein: coal, iron ore, fertile soil, trees, crops, etc. Land has always been considered a scarce resource that must be managed. Robots will affect the value of land, but not the underlying principle that to some degree, scarcity will always exist.

Timberlands and cornfields can be harvested and replanted in a sustainable manner if proper land use principles are applied. They are thus considered renewable. Fossil fuels, like oil and natural gas, are in theory renewable. They were created from the natural processes of the earth (land) and, if given enough geological time, they will form. Unfortunately for us, the renewal process for fossil fuels is eons longer than the human lifespan, making them nonrenewable for our purposes.

Technological advances will aid in the efficient harvest and use of the land's natural resources. For example, in the past decade, the petroleum industry's use of fracking and directional drilling techniques has resulted in the US doubling its production of oil, something previously thought to be impossible.

Technology and automation can be employed to better utilize the bounty of the land, but those resources will remain scarce, just as they were in the past. Regardless of the proliferation of robots, the economic principle of scarce natural resources will apply in the future.

Labor: Out of the Equation

It is my position that labor will no longer be considered a function of production because, in reality, it never was. Labor has been misunderstood because of our limited understanding of automation. In the past, human intervention was a necessary element of employing capital (tools). The caveman had to aim and throw the spear. The machinist had to operate the lathe. Capital could not operate in isolation from human effort.

Technology has advanced to the degree where automation can function with limited or no human input. Autonomous cars do not need drivers; 3-D printers do not need machinists. Labor never really existed; it was simply a misapplied term to describe the inefficiency of tools. Labor (human effort) was required because the tool could not operate by itself.

The robots are coming to take the jobs that have used misallocated human talent.

Entrepreneurship, Creative Human Touch

I believe that production consists of only three functions: land, capital, and entrepreneurship. The entrepreneur uses uniquely human abilities to create tools (capital) that organize natural resources (land) into valuable products and services. These products and services have value only to the extent that they satisfy the needs and wants of other humans.

People will be relevant and successful in the age of automation to the extent that they function as entrepreneurs, a uniquely human trait that cannot be mimicked by a robot.

The robots are coming to replace inefficiency, but they will not be eliminating the need for quality human interaction. On the contrary, technology will enhance our ability to engage with others. As you consider ways to monetize your human touch, look to old-fashioned methods that can be enhanced with modern technology.

My mother lived in or near the same community all her life. She maintained relationships with childhood friends because she interacted with them on a regular basis. I left home at age 18 and rarely kept up with old acquaintances as I traveled and lived around the world. My children are just as mobile as me; however, like my mother, they maintain old relationships because of social media.

My mother would smile and wave to an old friend in the grocery store. My kids "like" a picture posted by a childhood friend on Facebook. In either case, the interaction is minimal but results in establishing long-lasting relationships.

Just as social media allows you to stay in touch with old friends, technology will enable you to offer exquisite personal service to your customer base, service that will highlight your unique human attributes and make you a preferred choice over a lifeless robot.

The type of service is less important than the personal style in which the service is offered. To illustrate the point, meet Dr. Jill and Handyman Jack, two distinctly different individuals that have used technology to reinvigorate old-fashioned business models.

House Calls with Dr. Jill

Dr. Jill left her stressful job as an emergency room doctor to establish a family medical practice based on the old-fashioned business model of making house calls. Because of technology, she rarely visits a patient's home in person; usually, she conducts her exams via Skype. She also monitors her patient's vital signs via wearable sensors, like Fitbit or wireless glucose monitors.

Dr. Jill's target market is upper-income senior citizens with mild forms of chronic illness, like hypertension, diabetes, and Crohn's disease. Generally, her patients lead very active lifestyles and have the income to afford her concierge-style service. With technology, she is able to constantly monitor their condition, whether they are vacationing in Europe or entertaining grandchildren at Disney.

As an emergency room doctor, she might have seen as many as 50 patients per shift. Their cases ranged from ingrown toenails to gunshot wounds. The job was interesting, but not fulfilling. She worked long hours and never felt like she was in control of her own destiny.

In her new role, she has only about 200 patients. They pay in cash, so she does not have to deal with insurance companies or government reimbursement. She knows each one by name and is in constant contact via Skype, phone, or text. Dr. Jill is highly compensated and in control of her own life. She is extremely happy.

Handyman Jack

Handyman Jack has no formal education, but he is good with his hands and extremely curious. After being laid off from a manufacturing job, he started doing odd jobs in nearby neighborhoods until something better came along.

His need for startup capital was minimal. His assets consisted of an old van, a tall ladder, and assorted tools. With a few ads on Craigslist and word of mouth, he was soon busy cleaning out gutters, shoveling mulch, and hanging Christmas lights. He was willing to do whatever people asked and, consequentially, he was extremely busy.

Handyman Jack realized that there was much work to be done, but not all of it was profitable. So rather than taking a shotgun approach, he began to strategically analyze his opportunities.

He first developed a database of prospective clients in the more affluent neighborhoods. This was done using free resources from the Internet. The spreadsheet was a free version of Google Sheets. The information was mostly compiled by searching street addresses.

Second, he began to monitor social media to see what types of problems homeowners where having in those affluent neighborhoods. When he detected a trend, he would learn how to easily address at least a portion of the problem. Each neighborhood had its own lifecycle, and similar issues arose around the same time. It did not matter what the problem was or which appliance was affected—furnace fans, air-conditioning compressors, or garage door opener chains—they all tended to wear out simultaneously. If Handyman Jack could solve one homeowner's problem, chances are they would refer him to multiple neighbors with the same issue.

Third, he always fixed more than just the job he was initially hired for. For example, if a homeowner hired Jack to repair a malfunctioning garage door, since he was there and already had his ladder out, he would offer to replace smoke detector batteries simply for the cost of the battery, which he bought in bulk from Costco. If he was hired to hang Christmas lights, he cleaned out the gutters at the same time. Customers loved his simple, honest service and called him back again and again.

Handyman Jack now has a list of about 1,000 active customers in nearby affluent neighborhoods that he services on a regular basis. He is expanding his business by installing sensors that monitor conditions around their homes so that he can proactively address issues before the homeowner is even aware of a problem. Handyman Jack now works half the hours and earns twice the income that he did when he was employed in a factory. He is extremely happy.

Action Plan

Take an accounting of your wealth with these questions.

1. I have access to the following:

   a. Capital

      1. Amount of money that can be invested to develop a business concept:

         _____

      2. Amount of money that can be invested in traditional asset classes (stocks, bonds, real estate): _____

      3. I own the following equipment or tools that could be used to develop a business concept:

         _____

         _____

         _____

   b. Land

      1. I own the following real estate that could be developed or rented to produce income:

         _____

         _____

         _____

      2. My real estate contains the following resources that could be harvested or mined to produce an income (timber, farmland, water rights, coal, etc.):

         _____

         _____

         _____

c. Labor

    1. I possess the following marketable skills that can be used to produce an income:

_____

_____

_____

    2. I can hire the following talent to work on my business concept:

_____

_____

_____

2. Taking the above into account, do I realistically have enough wealth to support my desired lifestyle?

a. YES: I will put the above resources to work to create streams of income.

b. NO: I need to continue working at my present job and save more to build up my wealth.

Chapter 10

MONETIZING YOUR HUMAN TOUCH

As defined previously, human touch is the ability to create physical products and services, as well as intangible emotions like kindness and love. Human touch can be thought of as entreprencurship from an economic standpoint, because it is the sole human input required for production.

The entrepreneur uses capital (tools) to organize land (natural resources) into products, services, and emotions. To be successful, you must learn to monetize your human touch (entrepreneurial effort) because production is the source of wealth.

You might not think of yourself as an entrepreneur because of the limited definition of an entrepreneurship as a business owner. For our purposes, think of entrepreneur in the broader economic sense: the human creativity required for production. So, whether you are self-employed or an employee, to stay ahead of automation, you must be a *creator*.

Applied Economics

Mr. Spock, the fictional *Star Trek* character, once said, "Computers make excellent and efficient servants, but I have no wish to serve under them." Since you are reading this book, you probably feel the same way. So, how do you become the master rather than the servant? Let's apply some basic economics.

Under the old economic theory of production, human labor was a key element. As previously stated, I believe this was a misunderstanding of the inefficiency of capital and a misapplication of human talent. Labor was a means of quantifying how well humans could perform machine-like tasks. Labor was dehumanizing because the value was placed on precision and repetition, not creativity. Today and in the future, automation is so productive that it requires little or no human input. It will be inefficient to employ human labor to perform a task that can be accomplished solely by a machine—the robots are coming for those jobs that misallocate human talent.

At the lowest economic level, an employee only benefits from his effort to the extent that that effort can be profitable to his employer. The employee is compensated with an hourly wage or salary to accomplish a task needed by the employer's enterprise. The worker is a cog in the wheel. His reward is purely based on the value the employer has for the worker's effort.

If an employee's compensation is $20 per hour, it is likely that he is generating his employer at least $100 per hour. That is not extortion; it is simple economics. If the employer did not profit from hiring the employee, why would he hire the employee in the first place? That is the cold, hard fact about commerce, and the climate will get much colder for people that perform routine tasks. As technology advances and automation replaces human labor, the employee's earning capacity will be further degraded.

The employee uses his employer's assets to supplement his own effort. These assets are the traditional functions of production that we previously discussed—land and capital. The employer's land might be in the form of a coal mine, oil field, or farm. Or, the land might have a proximity value based on its access to markets, like downtown Manhattan or along the Mississippi River.

The employer's capital is in the form of tools—things like computer networks, equipment, infrastructure, and instruments—whatever items the employee needs to help him accomplish his job.

The employee is only compensated for his individual effort. The employer receives compensation from the successful implementation of his entrepreneurship:

- His own effort

- The effort of the employee

- Rent on his land

- Return on the use of his capital

The employee has one source of compensation, while the employer has four. Who do you think is more likely to build wealth?

Employees can rarely become financially independent because they are only paid on their individual efforts. Those efforts rarely produce large amounts of compensation. The exception is the employee with extraordinary skills, such as a neurologist (highly skilled professional) or a quarterback (highly skilled athlete or celebrity). Unless you are in one of these uniquely qualified fields, the opportunity to build wealth based solely on your paycheck is slim to none.

This compensation trap has always been the case, and it is one of the reasons that throughout history, only about 5 percent of the population has been able to achieve financial independence. Automation will increase income inequality because employers will invest in robots as capital equipment to replace less efficient workers. Employees will go from having low-paying jobs to having no paying jobs.

Commuter Thought Experiment

No one can know exactly which jobs will remain and which will be replaced by automation. To further complicate things, the very act of automation will create entirely new categories of jobs that have never existed.

Consider the impact of autonomous vehicles. At some point in the future, it is very likely that autonomous vehicles will reduce or eliminate the need for careers like truck driving. What is unknown is what types of new occupations will be enabled.

Let's envision the future through a simple thought experiment. Assume that autonomous vehicles exist and are in common usage, at least by the elite early adopters of the technology. Imagine a highly compensated corporate executive that has a one-way 45-minute commute to work. Previously, she had driven herself to work because it was not worth the expense to hire a limousine and chauffeur. The cost of the limousine and driver would not have been offset by the productivity improvement.

Sitting in the back of the limo would only make her slightly more productive than driving herself because while driving, she was capable of multitasking. She could drive and carry on a number of business functions beyond the normal phone conversation due to the advancements in voice recognition software. "Siri do this...Siri do that."

Here is where the thought experiment gets interesting. What if our executive can use the combined 90 minutes she commutes each day for purposes beyond basic business communication? If she can use that commute time to accomplish other tasks, then she gains over 390 hours per year. That is equal to over 48 eight-hour work days, or nearly 10 weeks of equivalent work time.

That productivity improvement is not only available to her, but to anyone that can afford to purchase an autonomous vehicle. Yes, trucking companies will use this to improve their profitability by firing truck drivers. But what other visionary entrepreneurs might adopt the technology?

How about service providers that normally would have rented brick and mortar office space? Imagine fleets of mobile offices using autonomous vehicles as their base of operation. Dentists, beauticians, therapists, whatever, all providing their services via mobile offices that capitalize on those newly liberated 390 hours of commute time.

How would it work? Monday morning, our executive is picked up by her attorney's autonomous vehicle, where she receives 45 minutes of legal advice on the way to work. That evening, she is picked up by her massage therapist. Tuesday, she rides to work in her personal trainer's mobile gym, and that evening, she commutes home with her piano teacher. This continues throughout the week until Friday evening, when she rides home in her beautician's mobile salon.

Sound insane? Maybe, but no more unrealistic than your current lifestyle would seem to my grandmother who was born in 1884. The point is that we cannot predict with certainty what type of jobs will exist in the future. What we can do is follow entrepreneurial wealth-building procedures that will work in any situation.

One-Concept Resume

What do you call an expert that claims to know exactly what jobs will be in demand in the future? A liar.

I am constantly asked what careers will be most profitable for the future. If I had to pick only one, broadly speaking, I would pick cybersecurity. That would include all aspects of network safety, from big data cloud computing to control of your home's thermostat. The field is wide, varied, and ripe for human talent well into the future.

Short of cybersecurity, every other field is at risk of being made redundant, as well as offering limitless opportunity. There will remain demand for *creative* physicians, plumbers, and attorneys. People will get sick, toilets will get clogged, and criminals will perpetrate violence. Creative people, at all levels, will be needed to solve these problems, just as they always have. The difference is that robots will easily perform routine tasks, so less competent (i.e., less creative) humans will be out of a job. Not to disparage our beloved public servants, but how many workers are really needed at the Department of Motor Vehicles?

If you want to remain employed, your resume will need to state only one theme: creativity. Employers of the future will not care where you went to school; their expert decision support system will contain all human knowledge. They will not care how fast you can accomplish a task; their robots work at lightning speed 24/7. Your potential employer will only be impressed with your ability to create new concepts. New, new, new.

The way to highlight your creative human touch skills will be by demonstrating items that you have created. Your resume, whether it is printed on a sheet of paper, attached to a LinkedIn profile, or posted as a video on YouTube, will only have impact if it demonstrates creative results.

The actual field is not important, because diverse markets will always exist. Those that will achieve will be able to combine emergent technology into useful things. The professional title may be called Carpenter, Cardiologist, Chef, Captain, or Cowboy—the function will be that of an integrator of technology into new products and services. Period.

Free Agent

Remember, the employee only has one source of income; the employer has four. I believe that in the future, those that remain employed will be rewarded with profit-like sharing plans that compensate workers for their overall contribution, not just an hourly wage or salary. This will occur not as a result of the benevolence of the employer, but to attract real talent, similar to the way a law firm offers a partnership to its top performers.

Technology's effect on employment will cut both ways. Productivity software will monitor each employee's performance down to the individual profit contribution, identifying who is least and most productive. Those with minimal contribution will be fired while the talented will be rewarded. The rainmakers will be given additional compensation not only to prevent them from moving to the competition, but more importantly, to keep them from starting their own enterprise.

Walt Disney needed thousands of employees and associates to bring his ideas to the cinema screen. This included illustrators, actors, cameramen, and darkroom employees, either employed directly by Disney or through partnerships with independent theater owners. Today, creative talent like Steven Spielberg can make due with less than 100 employees at DreamWorks; in the future, he will need even less.

UNICORNS & NIKE FACTORIES

Pandering politicians like to pontificate about bringing "good" jobs back to America. As previously discussed, if and when those manufacturing jobs return, they will employ robots, not people. This is nothing new. The best practices of highly successful companies have always been to minimize labor costs. That practice will never change. Technology will only enable it to occur at a faster rate.

Highly profitable manufacturing jobs have only existed to the extent that they continuously produce new products. This is due to the economic principle of diminishing marginal returns, whereby eventually optimization fails to increase output. Thus, the value is not in the factory infrastructure, but rather the creative process that feeds the assembly line with innovative new products to manufacture.

The mythological unicorn is much like a Nike factory, to the extent that neither exist. Nike sells $30 billion in athletic footwear and related products. Nike owns *no* factories. All manufacturing is farmed out to contract factories, mostly in Asia. A shoe made in Malaysia for $2.50 might be sold to US consumers for $100. The value is not in the manufacturing process. The value is in the design and marketing that creates emotional bonds with loyal consumers. That value emanates from the human touch skills that are employed within the walls of the Nike campus in Beaverton, Oregon.

Action Plan

1. I have created the following:

   a. Products or services (through current employer or personal entrepreneurial efforts):

   _____

   _____

   _____

   b. Artistic (portfolio of drawings, songs, poetry, etc.):

   _____

   _____

   _____

   c. Electronic or digital content (programs, apps, websites, etc.)

   _____

   _____

   _____

   d. Hobbies or interests (woodworking, sewing, landscape design, etc.):

   _____

   _____

   _____

   e. Social (started organizations, clubs, social media groups, YouTube channel, etc.):

   _____

   _____

   _____

2. From the above items, provide examples of market acceptance (sales, subscribers, downloads, etc.):

_____

_____

_____

3. I will incorporate examples from the above into my resume to promote evidence of creativity to prospective employers.

Chapter 11

FUTURE CAREER OPPORTUNITIES

So, what are the best career opportunities for the future? The future is unpredictable, and it would be naive to believe that a reliable list of the most resilient jobs can be conjured up. Uncertainty about technological advances makes it impossible to provide a specific list, but I will provide some general themes.

Throughout this chapter, I will use examples that involve medical professionals or medical devices. This is not to imply that doctors will be disproportionately affected by automation, but it is simply as a point of illustration. As you read these examples, try to apply the circumstance to your own situation.

White Collar Vs. Blue Collar

Before you totally discount some less prestigious jobs, think back to the examples of Dr. Jill and Handyman Jack. Those are two businesses that can work within the limits of today's economy. Their simple business models are framed around delivery of a concierge service. Roughly stated, they make house calls. The services they offer are quite different. Their education and accreditation level are polar opposites. She has a professional medical degree and is licensed by the state; he has no formal education but possesses mechanical skills. In spite of the education gap, both are highly compensated and extremely happy in their chosen fields.

The divergence between Dr. Jill and Handyman Jack illustrates the career possibilities for future self-employed entrepreneurs that eliminate working for the corporate middleman. For the self-employed, there is generally a smaller spread between the earnings of the white collar professional and blue collar craftsman. Dr. Jill may have a more prestigious career, but her lifetime earnings capacity is not much better than the handyman with no formal education. The doctor had to spend decades of her life and a small fortune to become accredited as a physician; the handyman had none of these opportunity costs and was able to start generating cash flow from day one.

The common denominator of their success is the ability to earn a living by providing a service that is based on their individual talents. They have learned to use technology to create a business model that makes it possible for them to monetize their unique human touch. Each earns a substantial income because they service a niche sector of the economy that is uniquely matched to their character traits. That is the essence of monetizing human touch. It also demonstrates an unlikely, yet potentially symbiotic relationship between the two service providers. They can both benefit from each other's service: Jill can use a handyman and Jack needs a family physician. Furthermore, since they are noncompeting and service an affluent clientele, it is likely that they could benefit from sharing customer lists. So, counterintuitively, a doctor and handyman have more in common than one might think.

I belabor the virtues of small-scale entrepreneurship because although millions will lose their jobs to a robot, creative risk-takers will find a profitable niche. The niche may not be a prestigious white collar profession, but it will provide an above-average income. This is a point worth emphasizing, because I believe our society has glamorized a college education to the detriment of hands-on vocational careers. As discussed previously, routine jobs will be the easiest to replace with robots. Redundancy will occur for *both* the white and blue collar worker if their job functions can be reduced to an algorithm. A formal education will not necessarily be the ticket to gainful employment. Less prestigious jobs will still be plentiful and profitable, as long as they are not easily replicated with a computer program. It will be hard to automate decentralized, nonroutine events like unclogging toilets, fixing broken equipment, and handling accidents. People will still be needed to respond to and fix these problems. Skilled service jobs will pay well. There just will not be as many as there are today.

Mirroring the Manufacturing Sector Trend

The overall future employment situation will be similar to what has already occurred in the US manufacturing sector. Manufacturing output has never been higher. Currently, the US produces products with a value in excess of $6 trillion. That places the US manufacturing sector larger than the entire gross domestic product of all nations other than China.

The problem for workers seeking manufacturing jobs is that all those "made in the USA" products are manufactured with less than 12 million people. Those employed in manufacturing jobs are less than 9 percent of the population, the lowest in history. The unfortunate reality is that good-paying manufacturing jobs still exist in the US; there are just not as many jobs as existed a decade ago.

The good news of US manufacturing is that those employed in the field have some of the best jobs in the world. The bad news is that only 12 million positions are available in a country populated by over 300 million people. This downward employment trend will accelerate as human workers (both blue and white collar) are replaced by robots in even greater numbers.

Think Outside the Corporate Structure

I keep stressing entrepreneurship because the majority of careers in the future will exist outside of the traditional corporate structure. This is because technology breeds productivity. Companies make capital investments to reduce expensive labor costs, thus reducing the aggregate number of corporate jobs available.

The employees at Facebook are 15 times more productive than the employees at General Motors. Each Facebook employee generates $1.25 million dollars in gross profits. Because of the huge profit potential, investor money is flocking to social media companies while automobile manufacturers are left floundering. The influx of investor capital will create more jobs at productive companies like Facebook. The problem for job seekers is that Facebook employs 200,000 less people than General Motors. This employment disparity is the undisputed trend of the future. This is why I believe that people will need to create their own employment opportunities.

Solve Problems

Self-employment need not be complicated, which is why I keep referencing the service businesses of Dr. Jill and Handyman Jack. Jack's customers do not care about credentials, they simply want household projects accomplished that they cannot or do not wish to do themselves. Jobs like Jack's are ones that are not likely to be done by a robot, nor can they be outsourced to lower wage factories overseas. Therefore, these manual type jobs will be able to command a premium wage. Also, many of these jobs occur as a result of a spontaneous event that requires immediate attention (i.e., a clogged toilet or lightning strike) so customers will be willing to pay top dollar to resolve the issue quickly. Consumers always pay for performance and convenience. The bottom line is that big corporations will not be able to solve every problem efficiently with automation, leaving many niche opportunities for entrepreneurs to harvest.

So, what is a good career for the future? Any job function that directly solves problems. The unfortunate situation for many professionals will be that technology will commoditize their expertise. This is the Achilles' heel that will sabotage many white-collar professions that are today considered prestigious. Consider the way something as innocuous as LegalZoom or TurboTax has lowered the price of legal and tax services.

For simple situations, a consumer can draft a will or calculate income tax for under $100. Technology has enabled consumers to find low-cost solutions that avoid the use of a high-cost professional. Good news for consumers, bad news for lawyers and accountants.

Robots in Medicine

Consumers seek the lowest cost solution that is appropriate to solve their problem. Automation and technology are rapidly creating products that bypass the need to consult with an expensive human professional. Continuing with a medical theme, consider the science fiction–like solutions that are commercially available today. Thirty years ago, it was impossible for an individual to map their DNA. Today, a DNA kit can be purchased from 23andMe for $99, no middleman physician or professional geneticist needed.

Kits and electronic sensors are available that can be used independently or combined with a smartphone app to diagnose or monitor many common medical conditions. There are numerous FDA-approved smartphone adapters currently on the market. AliveCor manufactures a device that records heart rate to detect atrial fibrillation; GluCase makes a blood glucose monitor for diabetics; ThermoDock's adapter records body temperature. Even more devices are under development or awaiting regulatory approval. Columbia University is developing a smartphone adapter microfluidic cassette that can diagnose severe infectious diseases like HIV and syphilis.

These devices, pending their approval, will revolutionize the way that consumers receive primary medical care. Consider the parent of a child with an earache. Currently, the child is taken to a doctor's office to confirm the presence of an ear infection. At best the office visit is time consuming, expensive, and probably exposes the child to numerous contagious diseases while sitting in the waiting room. All this inconvenience for a diagnosis that is relatively simple to perform.

CellScope is a company that manufactures a conical adapter that snaps over the camera lens of a smartphone. The device transforms the phone into an otoscope, allowing a parent to video their child's inner ear canal. The video can be emailed to a physician for a remote diagnosis. No need for physically visiting the doctor's office.

The logical progression of consumer-level medical diagnostics will be the complete bypassing of the human doctor's analysis. For example, in the case of CellScope's device, rather than having a physician view the video, why not use recognition software? The hurdle to adopting this type of diagnostics is not limited to technological feasibility but rather by regulatory and legal constraints. Kits and sensors can easily replace the services of many physicians and lab technicians. In complex circumstances where human oversight is needed or to control the distribution of pharmaceuticals, a less costly medical professional, like a nurse practitioner or physician's assistant, can be used. The regulatory and institutional forces currently limiting do-it-yourself (DIY) medical diagnostics will eventually moderate if for no other reason than the need to reduce medical costs.

Reducing the cost of medical services will be driven by both developed and emerging economies. Developed economies, like the United States, Western Europe, and Japan, have medical entitlement programs that cannot be funded because of lopsided demographics. The retired older population receiving benefits exceeds the number of young workers being taxed. To support the system, medical costs will need to be reduced to prevent the entitlement system from going bankrupt. Allowing robot-type diagnosis that fully or partially limits the need for an expensive human physician will be a viable cost-cutting method.

Robo-Diagnostics in Emerging Economies

Ironically, the initial demand for robo-medical services might be driven by emerging economies rather than mature markets, like the US or Europe. This path to technological implementation would be similar to the way that mobile phone technology has spread in less developed areas of Asia and Africa. These less wealthy nations lacked modern communication systems because they could not afford to install expensive copper wire telephone infrastructure. The problem was solved with the advent of mobile phone technology, which provided an affordable communication network because cell phone towers are scalable and much less expensive than hanging copper wire from telephone poles. Landline telephone installations in Asia never reached consumer participation levels seen in the US or Europe. However, mobile phone use is growing exponentially. There are over 1 billion mobile phone owners in China and over half a billion in India, far exceeding the 250 million users in the United States.

Developing economies leapfrogged over the copper telecommunication networks much like I believe they will bypass Western medical institutions. A village in a developing nation may not have the resources to staff a hospital or even employ a full-time physician, but the community could afford to jointly own a smartphone diagnostic device. This device would provide inexpensive diagnostics that would drastically improve public health issues, like combating contaminated water and identifying contagious disease. Problems can then be easily solved with inexpensive chemicals or pharmaceuticals once they have been properly diagnosed.

Since many of these nations do not have established institutions or regulations prohibiting robo-diagnostics, the adoption of this new technology might occur first in emerging economies. The devices are relatively inexpensive to own and require minimal training to operate.

Pharmaceutical manufacturers would likely subsidize the cost of the diagnostic device to create a sales channel for distribution of their products. The device would be a small one-time cost to the pharmaceutical company, but it would generate continuous repeat sales of consumable drugs.

The large consumer base in developing nations would provide incentive for medical device manufacturers to produce inexpensive user-friendly devices, similar to what has occurred with mobile phones. Consumers in developed nations would take note of the successful use of these "unauthorized" devices in other countries and thus enact legislation to legalize them.

THE AUTOMATED ANESTHESIOLOGIST

In 2013, the Food and Drug Administration (FDA) approved the use of Johnson & Johnson's (J&J) automated sedation system, Sedasys. It was initially authorized for minimally invasive procedures like colonoscopies. Sedasys offered an extreme value proposition with procedural costs of $150 to $200, compared to a human doctor's fee of nearly $2,000. Anesthesiologists are highly compensated for their skill, something in the neighborhood of $400,000 per year.

Sedation is a complex procedure; however, its core function is a rule-based logic decision biofeedback loop.

- Patient's heart rate is decreasing below threshold value:

  - Decrease sedation dosage

- Patient's heart rate is stable above threshold value:

  - Maintain sedation dosage

This analysis takes place in an anesthesiologist's mind. It is complex, yet routine. This type of logic can be programed into an algorithm, whereby a device controlled by a simple microprocessor can determine the amount of anesthesia to administer to a patient. Fast computers and advanced sensors have reduced the administration of anesthetics to a repetitive task than can mimic the functions of a human anesthesiologist.

As you might imagine, the device received a cold reception from the medical profession, especially the American Society of Anesthesiologists. Due to poor sales, in 2016, J&J pulled the plug on the automated anesthesiologist.

The argument made against the device was an issue of "patient safety." It was rationalized that since some people can react to sedation in a sudden and unpredictable manner, patient care was best left in the hands of a certified human anesthesiologist. Additionally, sedation is not only about the administration of drugs, but also the physical management of breathing and the airway.

I suspect there are logical arguments on both sides of the issue. Although Sedasys has lost the first battle, the war for automation of the operating room is far from over. The American Medical Association and other doctor's groups will oppose robots much like the United Automobile Works and Rust Belt labor unions have fought industrial automation. Ultimately, resistance will be futile.

Overcoming Institutional Resistance

Seem impossible? New technologies can rapidly be implemented once legislation and institutions prohibiting their use are eliminated. In fact, legislation need not be eliminated if consumers choose to ignore the law. Uber is illegal in many jurisdictions because of arcane laws favoring taxi cabs over private for-hire drivers. Yet, in 2016, Uber will book over $5 billion for its drivers.

Marijuana use is another example. Under US federal law, the use, possession, sale, cultivation, and transportation of marijuana is illegal. Yet 29 states have legalized cannabis in at least some form. In seven of those states, it is even legal for recreational use. So, at the same time that marijuana was illegal under US federal law, in 2016 the state-sanctioned marijuana industry had legal sales exceeding $7 billion. It was the fastest growing sector of the US economy. As Victor Hugo said, "No army can stop an idea whose time has come."

Established institutions always fight change. It is the purpose of their existence to maintain the status quo to protect their constituents. However, technology creates an opposing force that eventually builds enough momentum to overcome institutional resistance. As discussed previously, expert decision support systems will make much of today's required education unnecessary. Industry groups, like the American Medical Association and the American Bar Association, will do their best to stall the inevitable. Ultimately, mediocre doctors and attorneys will face stiff competition from automated diagnostic systems that can quickly identify an obscure disease or recall an ambiguous legal precedent. The knowledge acquired through decades of professional training will become commonplace with the use of a simple smartphone app. Competent professionals that survive will be those like Dr. Jill that can carve out a niche combining personal service, medical knowledge, technology, and business acumen.

EMBRACING INNOVATION

Highly educated professionals should not despair about their futures, provided that they are willing to use their education to enhance their innate human touch. Think of a world-renowned medical institution, and the Mayo Clinic will most likely come to mind. Many people might be surprised to know that the Mayo Clinic's main campus is not located in a posh metropolitan area, like Manhattan or Beverly Hills. Quite the contrary, the Mayo Clinic originates from Dr. William W. Mayo's humble medical practice, which was established after the Civil War in the remote, frigid town of Rochester, Minnesota.

Mayo's career was transformed during the societal and technological revolution that occurred during the mid to latter part of the nineteenth century. He was born in England and formally educated as a chemist. Mayo then immigrated to the US and eventually migrated westward. Like the people of his time, his career path was wide and varied. Primarily a pharmacist and physician, he was also employed as a census taker, farmer, ferry operator, justice of the peace, mayor, military officer, newspaper publisher, state senator, steamboat hand, and tailor—diverse experiences that no doubt contributed to his practical approach to practicing outcome-based medicine.

Mayo, along with his two sons, developed a medical practice that was quick to implement pioneering technologies, like the microscope and surgical antiseptics. Readily adopting cutting-edge technologies transformed the small backwater community of Rochester into a magnet for patients as well as top-notch medical talent. The clinic's accomplishments are legendary, including a Nobel Prize in Medicine for discovery of cortisone.

The success of the Mayo Clinic can be summed up in its heritage of adopting technology over conventional practices and constructing a team of like-minded, innovative doctors, two attributes which we all should try to emulate.

Action Plan

As you ponder the viability of future careers or business opportunities, it is important for you to ground yourself in *reality* by considering your earnings potential. Even if you are an employee with no intention to go into business for yourself, this exercise is important for you to gauge your employee value. Fill in the below information based on a real or imagined product or service that you could create.

1. I want to earn a total compensation of $_____ per year.

2. My product/service can sell for $_____ per unit.

3. My fixed cost per unit is $_____.

4. My variable cost per unit is $_____.

5. My total cost per unit is $_____. (#3+#4)

6. My profit per unit is $_____. (#2–#5)

7. I can produce and sell _____ units per week.

8. My total yearly compensation would be $_____. (#6×#7×52)

9. Are you being realistic?

    a. Is #7 attainable?

    b. Is #1 greater than #8?

Seem too simplistic? Perhaps, but as a late-blooming entrepreneur, I used a similar rudimentary process to map out my departure from a secure corporate job. The process reassured me that my assumptions were within the realm of reality. Grounded with confidence, I pursued my goal of self-employment in a field where I had no documented work experience. In less than three years, I had exceeded my previous six-figure corporate income.

PART THREE
Saving

Chapter 12

THINK LIKE
A SAVER

The replacement of human labor by robots will cause an economic shift like never before witnessed. Great fortunes will be both created and destroyed. For you to profit, you must be knowledgeable about markets and agile in trading strategies. The impact of robots will create extreme market volatility as old institutions are replaced.

In the remaining chapters, I will discuss a wide variety of economic topics that will be pertinent when automation disrupts the economy. To this point, you have learned to think like a human and an entrepreneur. Now you must learn to think in financial terms, like a saver, and then an investor. This is a very important concept and one overlooked by most experts. But the logic is straightforward: if earnings diminish because automation drives the cost of labor down, then deriving income from investments becomes more critical.

The following chapters will cover an extensive list of economic topics discussed in the light of a robotic future. Keep in mind that not all investments are of equal weight or are appropriate for everyone. The quality of investment opportunities will vary with market trends and one's ability to identify value, a skill that is required to monetize one's unique human touch.

The intent is to cover a wide range of investment possibilities while providing commentary to help develop your critical-thinking skills. The novice investor will be exposed to new concepts for further study and the more experienced investor will have strategies to pursue.

The following information should be important to you, even if you are not interested in investing. The market insight provided will help you recognize which sectors of the economy will be favored by automation and, thus, where to seek employment or entrepreneurial opportunities. Also, it is my premise that as the cost of labor becomes insignificant, wages will naturally decrease, thus increasing the need for income derived from investments and entrepreneurial endeavors.

Economic Reality

In a previous chapter, I discussed the difference in compensation between an employee and an employer. The employee is only compensated for his individual effort, while the employer profits from the success of the overall enterprise:

- His own effort

- The effort of his employees

- Rent on his land

- Return on the use of his capital

The common employee can rarely earn enough from wages to become financially independent. His skill level is usually not competitive enough to command a premium wage. Therefore, to build wealth he must become an owner, or partial owner, of a successful enterprise.

Ownership comes in many forms:

- Direct ownership with operating responsibility (e.g., self-employed beautician, electrician, or family practice physician).

- Direct ownership without operating responsibility (e.g., franchise owner that employs management and staff).

- Ownership of a rent- or dividend-producing asset (e.g., real estate or stock investor).

The first two items are related to the topics discussed in previous chapters about entrepreneurship and methods of developing your unique human touch. In the following sections, I will focus on the third topic, investment opportunities that are likely to be profitable in a world dominated by automation.

Own Appreciating Assets

Investing is complex, yet the underlying principle is simple: own appreciating assets. This is obviously easier said than done. Previously, I emphasized that you should *think like an entrepreneur, not an employee* and *like a human, not a machine*. Likewise, you should *think like a saver, not a consumer*.

Consumers purchase things for need and want. Savers make purchases for preservation of capital and future returns. In other words, savers put their hard-earned income into things (i.e., assets) that are likely to be worth more in the future, or at least not less. Saving requires the discipline to postpone gratification and the discernment to identify value, two key principles that will be rewarded in the likely deflationary cycle created by rampant automation.

A consumer buys a house because they like the yard or the kitchen. A saver buys a house because the neighborhood is economically trending upward and the house will likely be worth more in the future. Consumers live paycheck to paycheck. Savers live financially independent.

Savers put their money to work for them by investing. Think of investing as growing assets, like a gardener would grow vegetables. The vegetables will grow based on many things, including sunlight, temperature, water, and soil fertility. A gardener can only control some of these conditions but cannot protect against uncertain events, like an early frost. So, the gardener must be both prudent and vigilant in planting, transplanting, fertilizing, and watering based on ever-changing weather conditions.

Successful investing follows a similar methodology. A prudent investor "plants" her money in various opportunities and then vigilantly monitors how they are "growing" given the prevailing market conditions. If her assets are growing (i.e., appreciating), she leaves them alone; if they are not flourishing, she prunes them by moving her money to a different opportunity.

The concept of actively monitoring and adjusting asset allocation is unfamiliar to many investors because of the popularity of dollar cost averaging, or the practice of consistently buying investments regardless of market conditions. The financial industry highly promotes this practice. I believe dollar cost averaging is more beneficial for financial firms than for the individual investor. Their emphasis is always on "buy" but rarely on "sell." Even more rarely do they ever advise doing nothing at all and simply remaining in a safe cash position. The financial industry profits when consumers are purchasing investments, not when money is safely sitting in a bank or money market account.

When robots disrupt financial markets, just as employees will lose jobs, noncompetitive companies will become obsolete. Entire industries will be made redundant. This impact will be devastating to investors that simply buy general market indexes rather than selectively purchase shares of companies that are likely to profit from automation.

THE DEFUNCT TELEPHONE DIRECTORY

Shifts in technology have always had devastating effects on old industries. Just twenty years ago, there was a thriving multibillion-dollar industry supported by the telephone directory. (Millennials and younger, you will find this hard to believe, but prior to the Internet, addresses and telephone numbers were only accessible through a telephone directory. Each metropolitan area had their own book that contained residential and business information.) There was a large industry dedicated to producing these mammoth volumes. Lumberjacks harvested trees to make the paper. Immense manufacturing mills produced the paper and inks. Printers, book binders, advertising salesmen, delivery people—literally millions of people were employed in this sector. Since the data was constantly changing, the volumes of books contained obsolete information and had to be constantly reprinted. It was not very efficient but it was a fabulous way for lots of people to make money. Where are those companies and employees today? Mostly out of business.

Investors that "bought and held" sectors of the economy that profited from the telephone directory lost money. That included a wide range of investments, from stock in dividend-paying corporations to ownership of timberlands. The vast majority of these investments were considered "blue chip," or safe places to put retirement money. These stable institutions that had been incorporated for over 100 years ultimately ended up being horrible investments because they were displaced by technology.

Observe Changing Market Conditions

Think back a few decades, prior to the Internet. There were some really smart people working in the paper industry. The sector paid high wages and attracted talented and innovative people. They had to be innovative because it was a very competitive industry, and the companies that prospered were always finding ways to do things more cost effectively.

Generally, the industry embraced technology, even to the extent that paper might be replaced by plastic. However, their vision was myopic and their focus was limited to thinking in terms of substrates. A substrate is a physical material that could be printed on for conveying information, like a book, magazine, candy bar wrapper, or brown box. Paper manufactures could see the digital information age coming and, so, to a large extent, they tried to adapt their machinery to producing packaging materials rather than just paper used for printing.

Their emphasis on substrates caused them to overlook many ancillary opportunities, particularly in the emerging information technologies. The Mead Corporation, manufacturer of the ubiquitous spiral bound notebook, is a textbook example of overlooking an emerging technology, on par with Kodak's failure to advance digital photography.

Mead was an extremely innovative company. Their intellectual property and patent filings were so extensive that in the late 1960s, they started an internal project to digitally search legal documents. The technology was so effective that in 1973, they made it commercially available to law firms. What became known as LexisNexis started out as a hand-keyed electronic archive of legal cases in Ohio and New York. It soon blossomed into a searchable record of all US Federal and State cases and, more importantly, a database of major news articles. It was revolutionary. Lawyers and journalists were early adopters and paid handsomely for the service.

LexisNexis was so profitable that in 1994, Mead was able to sell it for an astonishing $1.5 billion dollars. A portion of those proceeds were used to help Mead diversify its manufacturing base by purchasing a plastic packaging company. The acquisition was seen as a twofold win. Mead was diversifying into plastic substrates and, more importantly, the plastics company held patents for an innovate design used for packaging CDs and DVDs.

Unfortunately, Mead's vision for the future was extremely shortsighted. Soon CDs and DVDs were in decline because software, music, and movies were simply downloaded. No packaging necessary.

The biggest irony of all is that LexisNexis was essentially the forerunner of a search engine technology that would later be developed by Google. The former Mead Corporation has since consolidated twice with two competitive paper companies; its parent WestRock has a market capitalization of

$12 billion. Alphabet (parent company of Google) has a market capitalization of $509 billion.

Technology shifts create winners and losers. Former industry giants can quickly become insignificant. A prudent investor will observe changing market conditions and adjust portfolio allocation accordingly.

Income Inequality

Automation has caused structural changes in the economy that I believe negate simple investment approaches that may have worked in the past. For example, a buy and hold, dollar cost averaging strategy consists of investing small amounts of money over long periods of time. This is the method that most people use, whereby a portion of each paycheck is contributed to purchase stock funds in an employer 401k plan. A buy and hold strategy was very profitable during the twentieth century when the US stock market returned an annual 10.4 percent[3] average rate of return. Buy and hold, dollar cost averaging strategies will be much less effective during periods of slow growth.

The Standard & Poor's 500 index (S&P500) is the financial industry benchmark for the market. The index represents the value of the US's largest 500 publicly traded companies. Its numeric representation is based on a "points" system that fluctuates with the stock prices of the 500 companies that comprise the index. The S&P500 closed at 1469 points on December 31, 1999. Nearly 17 years later, on October 31, 2016, it closed at 2126 points. Including dividends, that is a total annual return of approximately 4.6 percent, far less than the 8 to 12 percent most investors have been lulled into expecting. This has occurred at the same time that US Gross Domestic Product (GDP) has stagnated at 2 percent. It is my belief that this stagnation is related and primarily due to automation encroaching on employment.

Robots drive down human labor costs. It is no coincidence that average inflation-adjusted income peaked in 1999. The general economy cannot

---

3 John C. Bogle, Vanguard Group Chairman, Nov 11, 2003, in a speech at Vanderbilt University, "A New Era for Corporate America, for Mutual Funds, and for Investors."

grow if consumer earnings stagnate. Without broad consumer partic-ipation, only niche segments of the economy grow. Income inequality concentrates wealth and thus, a small number of investment opportu-nities rise while the majority stagnate or decline. The robots are coming, and income inequality will become significantly worse.

FACEBOOK MEDIA TAKEOVER

The success of Facebook is an example of how income inequality occurs due to a shift in technology. Concentration of wealth is not evil, nor is it a result of conspiracy. It is simply a result of market conditions converging and favor-ing a few individuals or sectors of the economy.

Throughout the twentieth century, most American cities had at least two predominant newspapers: Generally, one leaning to the left of the political ideology spectrum, and the other leaning to the right. There were numerous pub-lications that catered to large ethnic populations in urban areas, as well. It was also common for large newspapers to publish two daily editions: a morning and evening edition.

Like the telephone directory, newspaper publishing was big business. It employed millions of people from all walks of life: editors, writers, salesman, printers, all the way down to local delivery boys, who were literally boys as young as 12 years old.

The Internet has devastated newspaper circulation. Industry employment peaked in 1989 and revenue peaked in 2005. The industry is rapidly consolidating with only a few large companies remaining, like *The New York Times (NYT)* and *The Wall Street Journal (WSJ)*. Unfortunately, the survivors are not prospering. The NYT's stock is down over 50 percent since 2007.

While the online "new media" is thriving with readership, earnings are being concentrated among only a few com-panies. Facebook produces about $25 billion in revenue with 12,000 employees. The NYT revenue is $1.5 billion with 5,000 employees. Facebook's employees are more than six times more productive. That is why Facebook's founder Mark Zuckerberg has a net worth in excess of $50 billion and is the sixth richest man in the world.

In spite of the fact that the super-rich will become richer, you should make every attempt to build your net worth by prudently saving. Technological changes will affect many things, but historic asset categories (classes) are likely to remain a source of wealth for the average person. Broadly speaking, there are four asset classes that have endured the test of time and are likely to survive well into the future:

- Real estate

- Debt instruments (bonds)

- Commodities

- Company ownership (stocks and ETFs)

Each of these asset classes will be discussed in a separate chapter. Real estate and bonds tend to be more stable long-term investments and thus are being discussed in the light of saving. Commodities and stocks, being more volatile, will be covered from the perspective of investing.

Action Plan

1. Am I saving enough?

    a. My employer offers a generous and reliable pension plan (for federal government workers, etc.):

    1. YES: I should be saving at least 10 percent of my income for retirement. Proceed to item b.

    2. NO: I should be saving at least 20 percent of my income for retirement. Proceed to item b.

    b. By age 65, the aggregate of my pension, social security, 4 percent return on my savings, and up to a 6 percent drawdown from my savings would produce enough to cover my expected cost of living:

1. YES: You are most likely saving adequately.

2. NO: Proceed to #2.

2. I realize there are no shortcuts to building wealth. Unless my expectation is to subsist on a government-sponsored GMI, I have to do at least one of the following:

a. Earn more.

b. Spend less.

Chapter 13

REAL ESTATE

When discussing asset classes, real estate is a good starting point because for most people, a home is by far the largest asset they own. Also, as mentioned previously, real estate is "land," and land is nothing more than natural resources. So land, in a broad sense, is the beachfront property in Malibu, the coal under the ground in West Virginia, and the oil under the ground in Texas.

From an investment asset class perspective, real estate and the structures built on it will be considered aboveground property, while natural resources will be categorized as commodities. In the back of my mind, I still like to think of these two separate asset classes as "land," because the underlying trends that affect their value are essentially the same. As you read below, you will see that much of the affects that technology will have on the value of real estate can equally be applied to commodities. In either case, I believe that widespread adoption of robots will create price volatility and, ultimately, reduce the value of established markets.

Historically, real estate ownership has also provided a viable return on investment. Since many a fortune has been made in land speculation, a positive bias exists at all ends of the economic spectrum. From the emigrant who invests her life savings to become an apartment landlord to the dentist who purchases his own office space, everyone sees the profit potential of owning real estate. Wall Street has capitalized on the near-universal desire to own real estate by creating real estate investment trusts (REIT), financial instruments that sell shares in real estate developments of all sorts: housing, recreational, medical, etc.

While I believe that real estate will continue to be a productive investment in the future, a few words of caution apply, number one being that most people spend too much of their income on their home. Ignore debt ratios that are recommended by real estate agents or loan officers; their goal is to receive a hefty commission. They are not incentivized to care about your financial well-being.

Rule of Thumb for Buying Real Estate

A good rule of thumb for building wealth is to not purchase a house that exceeds three times annual income. For example, a household earning $100,000 should not purchase a house for more than $300,000. Many people would consider this advice unrealistic because they live in areas with high costs of living. Keep in mind, this advice is offered in regard to *building wealth*, not keeping up with the Jones or winning popularity contests.

While it is true that this formula is not achievable in many high-cost metropolitan areas, it is equally true that people living in those areas are not rapidly building wealth. Hence, the term "house poor," and the force driving many Americans to flee high-cost areas. Residents of New Jersey, New York, California, and Illinois are moving in droves to lower-cost states like the Carolinas, Tennessee, and Texas.

Consumers think in terms of income; investors think in terms of net worth. San Francisco leads the nation with the highest average compensation for technology workers. In 2016, average technology firm income topped $114,000 per year. However, average home prices in Silicon Valley exceeded $1 million. That is an income multiplier of nine times, far exceeding my recommendation of three times.

In spite of their high salaries, Silicon Valley workers are less likely to build wealth than the average American due to the high cost of housing. Just 80 miles away from the Bay Area, Sacramento's average home price is $280,000, making home ownership in Sacramento much more affordable for professionals like mechanical engineers, physical therapists, and transportation managers, who have an annual income of around $93,000.

This exodus out of high-cost areas is a microcosm of what I believe the future holds for migration patterns in general and why I see a red flag for real estate investing. While I believe that real estate will still be a viable future investment opportunity, I believe the rates of return will be lower than at any time in past history for the reasons discussed below.

Increase in Habitable Areas

In the distant past, people needed to live near sources of energy or transportation, generally large bodies of water. People naturally built communities along river banks or near coastal areas. Cities were initially established in habitable areas, like London or Chicago. As commerce developed, these places became great urban areas. As populations grew, these once-desirable places became overcrowded and overpriced. Residents remain because economic and social institutions provide access to jobs, quality schools, and cultural activities. As technology increases and robots become ubiquitous, expensive and crowded cities will become less attractive places to live because alternative enclaves will be established. Technology and automation will allow us to live anywhere.

Access to quality services, like education or medical care, will not be geographically dependent. Creative entrepreneurs that have learned to harness their human touch will not be tied down to a job in a specific location. People will be free to travel and live virtually anywhere.

While this will be good news for people with a sense of adventure and willing to migrate, it might be bad news for real estate prices in established areas. Connecting the dots to see how this could affect land prices is not too radical. Consider the migration of northerners flocking to southern states following World War II. Southern states offered many benefits over the North, including lower cost of living, lower tax rates, and warmer weather. So why did mass migration not take place prior to around 1950? Because of environmental factors. It was just too darn hot to live in the southern states during the summer. Although air conditioning was invented in 1902, it did not become commercially available until the 1930s. Like the adoption of most technologies, it was first utilized in business or industrial applications, then it became popular among the wealthy. It was not until around 1950 that middle-class homeowners and small businesses could afford residential AC window units.

Uncoincidentally, that is when the great exodus of northerners began migrating south. In 1950, New York State had about 15 million residents, five times that of Florida. That decade, southern states' populations began explosive "hockey stick" growth. By 2016, both New York and Florida had approximately 19 million residents. While New York's growth appears to be topping out, Florida is still growing at nearly 2 percent a year.

Technology allows mankind to control the environment. Air conditioning, among other things, made worthless Florida swampland inhabitable, significantly raising property values in the South while having a moderating effect on land prices in the North. This is an example of technology's influence on the inflationary/deflationary tug-of-war. We can expect emerging technologies to have similar effects on real estate prices in the future.

JAPANESE ROBOT FARM

In 2017, Japanese vegetable grower Spread will open the world's first large-scale robotic farm near Kyoto, Japan. The robotic farm factory will be based on a prototype indoor farm that Spread opened in 2007. Initially, the farm will grow lettuce in a monocrop system with the capacity to grow 219 million heads per year. The indoor farm will utilize soil-less growing beds stacked in vertical racks that can easily be tended by robots. Environmental conditions will be controlled using artificial lighting and heating; nutrients will be fed through a hydroponic watering system.

No pesticides will be used. The stable environment will decrease growing time by 30 percent and the use of robotics for planting and harvesting will reduce the employee count by half.

Indoor Farming and Land Usage

Light-emitting diode (LED) technology is revolutionizing the potential for indoor farming by providing inexpensive energy-efficient growing lamps. Previously, the cost of operating growing lamps was too expensive for all but the most high-dollar cash crops. Indoor farming was limited to growing products like luxurious orchids or illegal marijuana.

Today, the cost of LED units has plummeted, while at the same time their energy efficiency and effectiveness have increased. Their impact will not be as significant as the emergence of the personal computer, but it is likely to have a major effect on energy usage and the niche application of indoor farming. While indoor farming might not appear to be as revolutionary as air conditioning, in the decades ahead, I expect it to have a significant effect on land usage and overall real estate prices.

LED lighting opens up the potential for sustainable indoor farming, which could have a significant impact on those living in crowded urban areas as well as remote or previously uninhabitable regions. Urban indoor farming could permit food to be grown in skyscraper-type farms on a massive industrial scale. Indoor food production would not necessarily be price competitive for crops produced on a large scale, like wheat and soybeans, but could drastically reduce the price of foods that require labor-intensive harvesting or special handling, like many vegetables, coffee, cacao (chocolate), etc.

LED lighting would be used in conjunction with robotic technologies to create high-density indoor farm factories. Growing bins stacked vertically would be monitored by microprocessors to control robotic systems that administer just the right amount of light, nutrients, and water. Robots could be used to work the entire growing lifecycle of the crops: planting, feeding, watering, weeding, harvesting, and final waste composting. The farming process is very routine, making it extremely easy for operating systems to be written with simple algorithms.

The benefits of indoor farming could be further magnified because the uncertainty of weather conditions would be eliminated and pest infestations would also be easier to control. The potential is limitless and possibly self-sustaining. Recycled, nutrient-rich water from fish-containing aquaponic tanks would feed vegetables going in soil-less wicking beds. The closed system would provide healthy organic vegetables as well as a protein source from the fish. Indoor robotic farms could be established anywhere there was a reliable energy source. The energy could be derived from natural gas wells in west Texas, solar panels in Arizona, windmills in Iowa, or nuclear power in Japan. Small-scale robotic greenhouses could be contained within every home.

Indoor food farms would operate much like today's high-technology manufacturing facilities, with very little human labor. The cost impact will be profound. The value curve should mimic that of the consumer electronics industry: constant introduction of revolutionary products characterized by increased features at lower cost.

The effect on value and deflation will reach far beyond just quality food prices. Indoor, locally produced foods will require less inputs. Less transportation will result in a lower demand for fuel, vehicles, and drivers. Less water usage will require lower demand on reservoirs and aquifers. Less human labor will dramatically reduce the need for unskilled migrant workers. Less land use will reduce the need and, therefore, price of arable farm and timberlands. Imagine the impact on expensive farmland in regions like California's Central Valley.

SEASTEADING

Technological advances and automation will allow experimentation with all types of alternative living conditions. Seasteading might be one of these ventures. Seasteading is the concept of establishing permanent living quarters on the open sea, outside the jurisdiction of national governments. The primary appeal is to establish an unregulated sovereign community. PayPal cofounder and billionaire Peter Thiel is among those that have invested in the concept.

These nautical free trade zones could be housed on a variety of platforms, from artificial islands to huge floating modular pods. Building permanent structures in a harsh ocean environment is an expensive endeavor. To minimize the learning curve, the Seasteading Institute has plans to begin construction of a floating artificial island off the cost of French Polynesia in 2017.

The prototype structure will consist of a few floating platforms, perhaps 50 yards wide. Each platform will be designed for 30 occupants to live and work in sustainable quarters. The initial pilot project will cost $30 million.

Is seasteading the open oceans feasible? Who knows? Thiel's PayPal cofounder Elon Musk thinks seasteading is all wet. Musk's vision is to establish a colony on Mars.

Whatever new communities eventually emerge, expect them to make more efficient use of land and resources, thus continuing deflationary pressure on real estate and commodities prices.

Migration to Remote Areas

I believe the impact on real estate pricing could reach well beyond farmland. As air conditioning impacted the migration to warmer climates, indoor farming might encourage similar migration trends to colder or harsher climates.

North Dakota is not one of the most inviting of US states. The frigid state's population is less than 1 million. However, migration patterns into the state have been increasing because of oil and natural gas production from the Bakken Formation. Energy production from the Bakken is placing

North Dakota in league with OPEC-producing members, like Nigeria, Angola, and Algeria.

Extremely low energy costs, especially of natural gas, can turn shale formations like the Bakken into magnates for industrial development. Good-paying human jobs will coexist with massive robotic factories. Throwing indoor farming into the mix would make cold regions like North Dakota all the more welcoming to human inhabitants. Energy-abundant shale formations could become large metropolitan areas, offering residents high compensation in a low-cost environment. The same effect would apply to other remote areas abundant in natural resources, like Alaska or Russia. The lure of wealth (high income + low cost of living) would compel many people to migrate there, as did the California and Klondike gold rushes of old.

The confluence of technology will make remote areas more appealing because all the comforting features of civilization will be readily available: the Internet to access communication, information, entertainment, and virtual reality; indoor farming for succulent foods; robotic and advanced manufacturing techniques like 3-D printing to create an on-demand abundance of products and services.

Not enough to attract the wealthy elite from their beachfront estates in Malibu? Perhaps not. But the lure of lower cost remote living is not constrained to frosty Fargo or Siberia. Affordable undeveloped beachfront property is infinitely abundant along coastal areas of Mexico, Central and South America, Indonesia, and the Pacific Islands. Other than abundance of fossil fuel energy, the same technological features that would make far-off Siberia appealing would equally apply to sunny beachfront properties.

The average one-bedroom apartment in Malibu rents for over $4,000 per month. A luxury condominium can be had in Panama for $2,000. For the cost of renting a Malibu one-bedroom apartment, you could rent an entire luxury home along with accompanying wait staff in Bali. Less-known destinations are far more affordable.

DECLINING BIRTH RATES

The era of global population expansion is coming to an end. World population will likely grow to nearly 10 billion people and then start to contract sometime after 2050. Large families are favored in rural agrarian-based societies. Fertility rates have consistently declined since 1970 as the global population has migrated to urban areas access to healthcare has improved.

Population stability is achieved with a birth rate of 2.10 children per woman. The aggregate global birth rate is estimated to be around 2.5. The only regions with substantial population growth rates are in sub-Saharan Africa and the Middle East.

Among industrialized countries like the US and Europe, populations are declining with an average birth rate under 2.00. France's birth rate is among the highest in Europe at 2.07. The US's rate is far below population stability at only 1.87.

Russia, China, Canada, Germany, Japan, and South Korea are all at or below 1.60. Without immigration, by 2050 these countries' populations will likely shrink at least 20 percent.

Less people mean less need for land use and, therefore, potentially lower real estate prices.

Innovative construction methods, such as prefabricated modular construction, form-in-place walls, or robotic assembly, will also impact the cost of building. The possibilities are endless but all point to similar deflationary trends:

- Less use of labor and land

- Use of functional materials (phase-change insulation, solar roof)

- Intelligent features (monitoring, security, entertainment)

- Innovative architectural designs to take advantage of new technologies and renewable energy

Such factors are likely to make current residential and commercial buildings obsolete. So, even if the price of the land in Manhattan maintains its value, the buildings on it may need to be replaced with more intelligent and desirable features.

Declining Real Estate Values

Other trends are not favorable for supporting high future real estate prices. Online retailers like Amazon are decimating brick and mortar stores. Telecommuting and similar job arrangements are reducing the need for commercial office space. Imagine a world with no US Post Offices because they have been replaced by more efficient services like UPS or FedEx, or where shoppers no longer go to the mall because they purchase most things online, or where students receive their education online rather than attending brick-and-mortar school buildings. What happens to the price of all those buildings and the real estate they sit on? The value declines.

Apps like Priceline and Airbnb are reducing the cost of hotel and vacation rentals, as well as creating fierce competition by reducing barriers to entry. Advanced distribution will surely make the need for antiquated property use like the US Postal Service obsolete. (USPS owns or leases some 340 million square feet of land.) Lack of millennial household formation, as well as the transient nature of employment, will likely cause more people to rent rather than purchase a home. Declining birth rates are reducing all aspects of land use.

Rising interest rates will also negatively impact real estate prices. Interest rates have been declining since 1980. In 2016, mortgage rates hit historic lows; however, rates began to rise after the election of Donald Trump. Real estate prices are inversely proportional to mortgage interest rates, because the more borrowers pay in interest, the less money they have to spend on a house. As rates gravitate up from historic lows, real estate prices will naturally fall.

If all the above trends are not enough to squash future real estate yields, rising taxes might be. Property taxes are on the rise and the trend is likely to swiftly escalate. Property taxes are among the only source of income for local governments, whose deficits are rising because of expansive services and underfunded pension plans. From Detroit to Seattle, property taxes are growing double digits with no relief in sight.

In Texas, the $2.39 billion Dallas Police and Fire Pension Fund is forecast to become insolvent by 2028. The mayor is concerned about the pension fund bankrupting the city, or property taxes having to be raised 130 percent to cover shortfalls. Such a confiscatory property tax hike would paralyze the Dallas real estate market.

Many cities are faced with huge legacy pension fund shortfalls. As they raise property taxes to cover the gaps, it will discourage residents from living there, further eroding the tax base. An economic brain drain will result as mobile entrepreneurs flee to lower cost tax havens.

Combining all these reasons with the long-term trend of moderate monetary inflation, I do not see real estate as a compelling long-term investment opportunity, certainly not to the degree it has been in the past 50 years. My recommendation is not for you to immediately dump your house, but, rather, to use caution. In the long term, I expect real estate will generate returns that moderately keep up with inflation.

Action Item

1. Should I own or rent?

    a. It is likely that I will move in the next 3 years: _____

    b. My job or income is not stable: _____

    c. I have little to no savings: _____

    d. A mortgage payment exceeds 35 percent of my pretax income: _____

2. I answered "YES" to more than one of the above:

   a. YES: Proceed to #3

   b. NO: I can probably afford to own a home

3. The general societal bias is to own a home, but based on the above answers, I might be financially better off if I rented rather than owned.

Chapter 14

DEBT INSTRUMENTS (BONDS)

Bonds, debentures, and other forms of debt instruments are popular among prudent long-term investors who are seeking stable income payments. While I would certainly agree that bonds can be an appropriate investment opportunity, I believe bond prices, like real estate, will be much less dependable in the future.

Bonds are heralded as "safe" investments, so much so that advisors frequently use a simpleton formula for calculating stock/bond ratios in retirement portfolios. The erroneous explanation is that since bonds are safe, as one ages, more of their portfolio should be invested in bonds. The basic formula states that an investor's age should dictate the percentage of bonds held. A 65-year-old investor would hold a portfolio that is 65 percent bonds and 35 percent stocks.

There are several flaws to this reasoning:

- First, bonds are only as safe as the stability of the guarantor. US federal government bonds are extremely safe, but bonds issues by states, cities, municipalities, or corporations, not so much. Unsophisticated investors tend to gravitate toward bonds paying the highest yields but neglect to account for risk. While bonds may be safer (better price stability) than stocks in general, a high-yielding bond issued by a sketchy corporation or municipality could be much more unstable than a quality blue chip stock.

- Second, most retail investors do not hold actual bonds but rather purchase shares of bond funds. Bond fund prices fluctuate based on interest rates (bond prices move inversely to interest rates). For the past 35 years, this has not caused a downside risk because bond prices have been increasing as yields have decreased. Presently, global interest rates are at historic lows, some issuing negative rates. Reason would dictate that at some point, rates will rise and, therefore, cause bond funds to lose value.

- Third, as discussed throughout this book, robots are going to have a profound impact on the economy. These effects will likely be devastating to old, established institutions. Uncompetitive companies and the local governments supported by their tax base could become insolvent, leaving their long-term bonds worthless.

Bond Rate Forecast

As far as an actual rate forecast, I think that investors hoping for a return to a ten-year US Treasuries interest rate of above 4 percent are going to be disappointed. I do not believe that overleveraged governments, corporations, and consumers can service existing debt above a 4 percent 10-year US Treasury interest rate. If this were to happen, spending would plummet and the already-weak economy would fall into recession.

The only way to get to higher interest rates without causing a recession (i.e., killing consumer spending) would be a forgiveness of the over $1 trillion of student loan debt and/or some type of national healthcare that reduced the burden of insurance premiums and deductibles on the middle class. Student loan debt and healthcare costs are consuming what would otherwise be the middle class's disposable income. I am not suggesting that these policies would be good for the economy long term, or that I support them. But if tax cuts and infrastructure spending fail to sustain GDP above 2 percent, then expect these and more "stimulus" spending initiatives to be proposed.

Historically, the 10-year interest rate is approximately equivalent to nominal GDP, meaning growth including inflation. Since the recession of 2008, many believe that central bank intervention has suppressed rates well below this historic level. It could also be that reported GDP growth and inflation are both too high. In any case, my forecast of sub-4 percent rates is derived from the premise that should rates "normalize" to the historic relationship with nominal GDP, then: 2 percent real GDP + 2 percent inflation = 4 percent 10 year US Treasury yield. That reflects the average GDP since 2000 and meets the Federal Reserve's inflation target.

Many would consider a 4 percent yield too low, given much higher rates over the previous 30 years. This gets back to the inflation/deflation tug-of-war. As long as inflation does not get ahead of deflation, then higher interest rates cannot likely be supported by our anemic economy. The double-digit inflation and corresponding interest rates of the 1970s and 80s was actually a historic anomaly, something the economy has been working off for the past 35 years. Ten year US Treasury yields varying between 2 to 4 percent were the norm from 1875 to 1965.[4]

Lower Barriers to Entry

Lack of capital investment is also a reason that interest rates have been lower for longer. If large corporations do not need to borrow massive amounts of money, then the cost of borrowing money (interest rates) remains low. It is a function of supply and demand. In the 1970s, if a large industrial company like General Motors or 3M expanded its business, it took large amounts of capital to re-tool and build new factories. Today, many technology companies, especially Internet firms, are totally debt free because they operate so much more efficiently than old Rust Belt companies. Google, Facebook, and Adobe have no or very little long-term

4 Global Financial Data, "USA 10-year Bond Constant Maturity Yield (IGUSA10D)," http://ritholtz.com/wp-content/uploads/2010/08/1790-Present.gif.

debt. Apple has over $200 billion in cash on its balance sheet and borrows money only because it is less expense to borrow than to pay tax on repatriated offshore money.

This is an important takeaway for entrepreneurial-minded people: Money is cheap, and likely to remain that way. Also, startup capital requirements are much less expensive than in the past. That means lower barriers to entry for entrepreneurs who want to enter a new market. There has never been a better time to start a business.

Another future trend that might act to depress interest rates is the rise of peer-to-peer lending and crowd sourcing. Both of these methods of raising capital circumvent the traditional banking industry. Interest rates charged by peer lending groups can be substantially lower than conventional loans, especially when compared to credit card rates.

I expect alternative lending programs to grow in popularity until the next major recession. At that time, if default rates skyrocket, peer creditors might find out they are not receiving an adequate risk-adjusted return. Uncollateralized loans are always risky and peer lenders should recall the Wall Street adage, "Past performance does not guarantee future results." Committing more than 10 percent of one's portfolio to peer lending is probably not prudent.

Rising Rates

A rapid rise in interest rates after a decade of Federal Reserve suppression could cause a dramatic decrease in bond prices, thus bursting the "bond bubble." The risk for this would rise significantly if rates were to exceed 4 percent, again supporting my forecast that 10-year US Treasury yields will not likely rise much beyond that level.

Should rates rise dramatically, I would not be surprised to see a collapse in the global bond market. Like the prior bubbles of tech, housing, and finance, a bond price collapse would set off a global recession, which would then provide an excellent entry point to reinvest in the general market. Until then, I recommend that bond holdings be limited to the most liquid, high-quality investment grade available.

TRUMP BOND BUST

Think US Treasury bonds are a safe investment? If you own a bond fund, better think again. The extremely popular iShares 20-year US Treasury Bond Fund (TLT) is a case in point. Its price is inversely proportional to interest rates, meaning that when interest rates rise, its price falls.

The price fluctuation can be very volatile, often surprising unaware investors to the downside. What happened after the November 2016 election of Donald Trump is a classic example of an abrupt drop in bond prices. From the week prior to the election to the week following, TLT lost almost 7.5 percent of its value. Over a month-and-a-half period, from its highest to lowest point, TLT lost over 12.5 percent. Calculated from the four-month pre-election high, the fund lost over 15.5 percent. Any investor who had bought into the fund at any time during the 10 months preceding the election would have lost money. In fact, purchasing TLT at any time in the two years preceding the election would have resulted in a post-election loss about 80 percent of the time.

Such abrupt moves to the downside are shocking to retired people who erroneously assume bond funds are "safe." Do not get caught off guard. Bond funds can, and often do, lose money.

Action Plan

1. I own bonds or bond funds that have the following in their title:

   a. High yield: _____

   b. Alternative: _____

   c. Non-investment grade: _____

   d. Investment grade below BBB (BB, B+, C, etc.): _____

2. Bonds that fit into the above categories might be considered "junk" with a significantly higher default rate than investment grade bonds. It is probably not prudent to have more than 10 percent of my portfolio invested in junk bonds.

PART FOUR
Investing

Chapter 15

THINK LIKE AN INVESTOR

Regardless of what type of asset class you invest in (real estate, stocks, etc.), always stick with quality and liquidity (meaning the asset can easily be bought or sold). Quality does not mean popularity or media coverage by talking heads. Quality means the ability to consistently increase earnings. Quality equates to value.

If a company does not have any earnings or if its earnings growth is declining, you probably do not want to own it. There are exceptions to this rule, like when a company is recovering from changing market conditions or if the broad economy is suffering from a recession. But even in those instances, at a minimum you want to see some light at the end of the profitability tunnel, indicating that the company is indeed headed for recovery.

Avoiding unprofitable companies means that you will inevitably miss out on owning innovative new startups. Considering the failure rate of new companies and emerging technologies, I consider this a worthy tradeoff. The vast majority of startups fail. I am willing to miss the opportunity of a few winners, so long as I preserve my capital by not losing on all the failures. Like the tortoise, slow and steady wins the wealth race. The speculative hare almost always finishes broke.

As an individual investor, you have no control over the management of the company or any insight into the inner workings. Investors are limited to identifying trends and then trying to construct trades in companies that are likely to be affected by those trends. Note: I did not say "favored" by the trend, the word I used was "affected." The affect could be positive or negative. Robots will benefit some companies while other companies that cannot adapt go bankrupt. Automation will devastate many established institutions, particularly those that derive their competitive advantage from some type of sanctioned monopoly status. We are witnessing this play out with Uber's fight against the protective operating medallions in the taxi cab industry, which limit business licenses to a few privileged companies.

The key point is to *think like an investor, not a speculator*. Avoid get-rich-quick schemes. Do not try to find a needle in a haystack or kiss frogs hoping for a prince. Invest in companies that have shown a propensity to earn money in the past *and* appear to be adaptable to changing markets and technological conditions. Do not buy "junk" corporate bonds or near-bankrupt municipal bonds just because they pay a high yield. Stick with quality and value.

How can you identify companies that have a management culture that encourages adaptive forward thinking? Most likely, they are the same companies that have consistently increased their earnings. It may appear as circular reasoning but in an uncertain world, consistent earnings are the best indicator of adaptive corporations. It is not fail-proof, but it is the best indicator available. To paraphrase Forrest Gump's mama, "Value is as value does."

Active Investing Is Not Speculation

Not all established institutions will fail. In fact, many will flourish, especially at the beginning of the robot takeover. More automation means less expensive human workers, thus, greater corporate profits. To take advantage of these productivity gains and to protect against broad market stagnation or catastrophic loss from a market crash, it is important to actively trade your portfolio. Active investing is not speculation, it is thoughtful behavior, like that of the prudent gardener who prunes back her plants. Likewise, the wise investor will need to watch market conditions and proactively adjust their asset allocation.

Active investing is the alternative investment approach to dollar cost averaging into a general index fund. If you feel uncertain or ill equipped to actively trade your investment portfolio, then by all means, you should not. However, at a minimum, you should read on and become familiar with such concepts. This is critically important because automation will affect the economy at a rapid pace. Today's Google and Amazon will ultimately be replaced by other, more innovative companies. It is the nature of a competitive marketplace.

While it is true that markets generally recover and revert to their averages, it is also true that upward movement can take a long time. The collapse in technology stocks during the dot-com bubble is a classic example.

The leading index for technology stocks is the National Association of Securities Dealers Automated Quotations (NASDAQ). During the irrational exuberance of the dot-com bubble in 2000, the NASDAQ hit an all-time record high. It was more than 15 years before the NASDAQ again surpassed that level. Since the gain was in nominal terms, meaning that the price was not inflation adjusted, investors that took a buy and hold approach after the collapse of the dot-com bubble have still not covered their loss. In fact, the NASDAQ would need to appreciate approximately another 35 percent from 2016 levels for dot-com bubble losses to be recovered in real terms (inflation adjusted).

Active investing (not speculating) means moving assets from depreciating or stagnant sectors to those that are appreciating, or at least keeping up with inflation. The strategy is to preserve capital (keep up with inflation) and seek a return. The emphasis of this approach is on the word "active," because markets are constantly changing.

How a Bubble Is Formed

Thinking like an investor will help you to *avoid* a catastrophic loss, which is often the fate of a speculator. Mania "bubbles" usually form when speculators bid up an asset's price beyond rational fundamental values. This can occur in real estate, stocks, or any asset class, including bonds. In recent years, the hype over 3-D printing formed a large bubble in that small industry sector. Prior to about 2011, no one outside of industrial manufacturing had ever heard of 3-D printing. It was not a new concept: The first patent for "stereo lithography" was issued in 1986 and 3D Systems was founded to industrialize the invention. The technology had great potential but was slow to catch on. Application equipment was primitive and useful materials were scarce. Aside from a few maker hobbyists and devotees of *Popular Mechanics*, the only people using 3-D printing were rapid prototype engineers.

Then came the hype, from *Ted Talks* to magazine covers; 3-D printing was fashionable. Talking head chatter started to ramp up in early 2012 and reached a zenith in May 2013. The industrial possibilities of 3-D printing suddenly went from obscure niche to mainstream, and its use was seen as a cure-all for every problem that plagued modernity.

3D Systems, the small, unprofitable company, became the darling child of Wall Street speculators. Almost overnight, 3D Systems rose from anonymity to technology acquisition kingpin. At the stock's peak in late 2013, 3D Systems' market capitalization was valued in the neighborhood of $10 billion.

Then the bubble burst. People started to come to the realization that although 3-D printing offered immense possibilities and the promise of astronomical profits, the time had not yet arrived. By the time the retail-investing public realized that the proverbial emperor was not wearing clothes, speculators were already fleeing the stock in droves. Two years after 3D Systems' stock peaked, the price fell by 90 percent, entirely wiping out the gains of the previous five years. A similar tale can be told about 3D Systems' major competitor, Stratasys.

The stock price collapse had nothing to do with the validity of the future potential of 3-D printing as an industry. Nor was it a negative commentary on the worthiness of 3D Systems as an industry leader. I strongly believe that 3-D printing will revolutionize manufacturing practices and have an enormously positive impact on virtually every sector of the economy. Furthermore, I believe that 3D Systems is positioned to take advantage of that trend and at some future date will be immensely profitable. But all that is in the future, possibly far in the future.

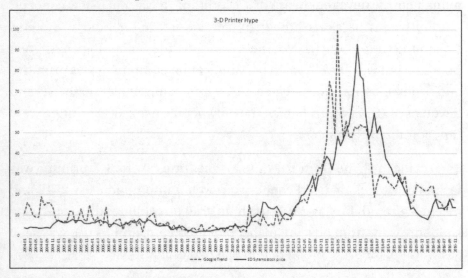

The above chart illustrates perfectly the pattern of an equity bubble. It is not unique to 3D Systems or the 3-D printing industry. Note how the stock price is closely correlated with media hype, in this case incidents of "3-D Printing" as tracked by Google Trends. This bubble pattern plays out time and again, across all asset classes. Whether the buying frenzy be in gold, real estate, or a company stock, this same general pattern prevails.

The mania pattern is characterized by astronomical price increases: It rises from the promise of enormous future profits and falls with a price collapse of at least 80 percent from the peak. In the end, cooler heads prevail and a few strong companies survive to become industry leaders.

Study history, and you will see that this pattern appeared at the introduction of any innovative technology: railroads in the mid to late 1800s; automobiles in the early 1900s; nuclear energy in the mid-1900s;

computers in the late 1900s; and the Internet in the early 2000s. Today, it is robots.

The lure of get-rich-quick speculation captures the popular culture and eventually ends abruptly with the fear of catastrophic loss. The mania has no relevance to the specific technology, it is simply a manifestation of human nature. That is why the bubble pattern is such a reliable phenomenon that reoccurs time and again. Technologies come and go, but human nature never changes.

Wait for this pattern to form. It always ends in an abrupt price collapse. A prudent investor will become familiar with it and use it as an early warning sign to avoid catastrophic loss.

Avoid the Bleeding Edge

How can an investor avoid a catastrophic bubble? There is no guaranteed safety net, but I recommend sticking with quality. Quality translates into solid earnings, not just increasing sales. Companies that are generating consistent profits are, by definition, not on the "bleeding" edge of technology. Bleeding-edge companies hemorrhage money during the early startup phase in an effort to gain market share, which is a very risky strategy. Ignoring these high-flying media-hyped companies is not exciting, but it can be advantageous to your net worth.

As outlined above, the stock performance of emerging innovative sectors of the economy always follow a predictable pattern:

1. A new technology emerges.

2. Companies rush in to gain an advantage in the new sector.

3. Media attention enthusiastically promotes the investment opportunity.

4. Unsophisticated "retail" investors rush in to greedily buy stock.

5. Stock prices rise to unjustifiable valuations, prices that will never be supported by realistic future earnings estimates.

6. The mania bubble bursts, with stock prices falling 80 percent from the high.

7. A few large companies survive and eventually dominate the sector.

This same bubble pattern plays out time and again, only varying with intensity. On a small scale, the bubble may form with just one company, whose stock price is being manipulated by unscrupulous promoters to orchestrate a "pump and dump" scheme. Or, it could be a colossal mania that creates a global financial crisis, like the 2000 dot-com bubble or the 2008 housing bubble.

As I have mentioned several times, from a career perspective, I believe cybersecurity is an excellent field for a young person to pursue. Likewise, fortunes will be made by investing in the cybersecurity sector; however, cybersecurity stocks fit the definition of a bleeding-edge technology that is likely to burst, resulting in more losers than winners.

The field is ripe with potential superstars like Palo Alto Networks, FireEye, and Proofpoint; however, despite the fawning media coverage that these companies receive, they are not profitable. The cybersecurity sector will undoubtedly produce huge profits, but for which firms? Will Palo Alto Networks dominate FireEye? It is anyone's guess.

Following the methodology of trying to avoid the bleeding edge, rather than flipping a coin between two unprofitable superstars, I would prefer to limit risk with a more diversified technology company. So, rather than risking it all with an unprofitable pure play in the cybersecurity industry, why not just dip your toe in the water by investing in a company that is in that space but has other sources of revenue—perhaps IBM, Cisco, or Symantec? While none of these stodgy companies offer the explosive growth opportunity of the younger upstarts, their reliable profits limit a catastrophic loss.

Also, a company with deep pockets, like Cisco or IBM, is likely to remain relevant in the cybersecurity space even if innovation does not originate from within. Cisco has some $70 billion in capital that could easily be deployed to buy the appropriate cybersecurity technology once a clear solution emerges. First to market does not always guarantee market share dominance. Recall that Apple did not invent the personal computer, MP3 player, music downloading, smartphone, or anything else.

Another argument for investing in an asset-rich company like IBM over a small and glamorous cash-poor type like CyberArk Software is that IBM has cash-producing infrastructure in place. From a global service network to sales and marketing teams, IBM's infrastructure is entrenched deep within the world economy. Legacy companies like IBM are also capable of organic innovation. IBM's Watson supercomputer might prove to be the mainspring of expert decision support systems.

Key Takeaways

Just as this book could not advise you on *exactly* the right career path for the future, neither can it make you a skilled investor. The point is to help you develop critical-thinking skills so that you can assess future situations and react accordingly, thus helping you to think like an investor, not a speculator. There are two key takeaways that reinforce why active investing will be so critical to your ability to build wealth in the future:

1. Markets fluctuate in broad generalized trends that last for extended periods of time: months, years, or even decades. The point is that an investor does not have to precisely predict the bottom or top of a market. The key is to identify a broad trend and then position assets to take advantage of it.

2. Gains increase dramatically *more* in a market recovery than they fall during a market downturn. This is a matter of simple mathematics, but it is overlooked by most people.

   A substantial percentage increase will always occur when a market recovers, even if the price just returns to the previous high. There is no magic, it is a simple mathematical reality:

a. Original price: $100

b. Price following a catastrophic decline of 50 percent: $50

c. Recovery to original price of $100 nets a 100 percent increase from the bottom.

Note that the loss on the way down was 50 percent, but the gain on the way back up was 100 percent. The lesson is that a buy and hold strategy would have caused an investor to suffer through the pains of waiting for the market to recover, and in the end, have no capital appreciate.

Active investing strategy would have resulted in a profit even if trades were executed at market midpoints.

Selling at $75, when the price had declined 25 percent (from $100) and purchasing at $65, when the price had recovered 30 percent (from $50) still results in a net profit, provided the price recovered above $65. A pro-active approach could be executed far from market top or bottom and still return a handsome profit.

Action Plan

Answer yes or no to the following questions.

1. I primarily make investment decisions based on:

   a. Tips from friends and family: _____

   b. Advice from investment shows on TV, radio, podcasts, blogs, etc.: _____

   c. My gut feeling: _____

   d. Financial advertisements: _____

2. I answered "YES" to any of the above:

   a. YES: I might be engaged in speculative behavior and should reconsider my approach to investing.

   b. NO: I will continue to be vigilant and seek out sound investment opportunities.

Chapter 16

COMMODITIES

Commodities are tradable raw materials or agricultural products, things like iron ore, gold, or coffee. Traditionally, investors have traded commodities for either short-term speculation or as a long-term hedge against inflation.

Short-Term Speculation

Short-term speculative trades take advantage of disruptions or anomalies in the supply chain that are caused by uncertainties like weather conditions. For example, a speculative trader may try to make a quick profit by investing in orange juice futures after a winter frost devastates Florida citrus groves.

Short-term speculative trades will always present opportunity to take advantage of uncertainties caused by weather, natural disaster, and war. However, speculative trades involve elevated amounts of risk that are not appropriate for preserving capital. The prudent investor would not use money saved for long-term goals like retirement for speculative commodity trading.

Long-Term Hedge Against Inflation

Investing in commodities as a long-term hedge against inflation can be prudent, with restrictions. Most investors fail to realize that over the long run, commodities generally only keep up with inflation. Investing in commodities can help *preserve* wealth, but not necessary increase it.

For example, if someone buried $10,000 in their backyard in 1970 and reclaimed it in 2016, the money's purchasing power would have declined by 84 percent, losing $8,400 of 1970 purchasing power.

To illustrate the loss of purchasing power, in 1970, $10,000 would have purchased a Cessna 150 two-passenger commuter airplane. In 2016, the same amount of money might be enough to buy a good used car.

Had the $10,000 been used to purchase 256 ounces of gold in 1970, the gold would have a value of over $300,000 in 2016; not enough to make you rich, but certainly enough to purchase a small commuter airplane.

Purchasing other commodities in 1970 would have preserved value less than gold. Silver would have a present value of less than $100,000. Oil's present value would be less than $90,000.

The point is that generally, a basket of commodities will keep up with inflation over time. Investing in one commodity may not, especially if the time period is less than a decade. As of 2016, the majority of investable commodities has declined in value over the past five years.

My opinion on the long-term performance of commodities is the same as my position on real estate. I believe the exponential increase in commodity prices that has occurred since World War II has already begun to dissipate. I expect commodities overall to keep up with inflation; however, the value of specific commodities could drop dramatically.

Inflation/Deflation Tug of War

The robots are coming to affect more than just the job market. Robotics will reduce the cost of almost everything, especially commodities. On an inflation-adjusted basis, with the exceptions of healthcare and education, many products and services have gotten cheaper. Copper is an industrial metal whose price is so highly correlated with the economy that it is often dubbed "Dr. Copper" because shifts in its price reflects movements in the economy better than prominent PhD economists can do.

So, what is copper telling us about inflation? Copper's average price in the early 1980s was approximately 95 cents per pound. In 2016, copper's price fluctuated around $2.25. So, over more than a 30-year period, copper's price is compounding at an annual inflation rate of approximately 2.5 percent, which is fairly consistent with the Federal Reserve's long-term goal of maintaining a 2 percent inflation rate. People worried about hyperinflation are consistently disappointed.

Inflation is constantly being challenged by deflationary forces. Most people do not notice the positive effects of deflation because they see the value of their money eroded by inflation. Inflation has several sources, but the primary reasons money loses value is because the central bank expands the money supply and the government runs large deficits.

The tug-of-war between inflationary and deflationary forces is important for investors to comprehend so that they do not become unduly pessimistic. As government debt escalates to unfathomable amounts, a sour mood has overtaken many people because they fear economic collapse is imminent. To protect against hyperinflation or a currency default, they purchase precious metals like gold. I do not believe economic collapse gloom-and-doom scenarios are likely, because the inflationary/deflationary forces tend to counterbalance each other.

The checks and balances will work something like this:

- Automation will result in vast unemployment.

- Those profiting from the effects of automation will be heavily taxed.

- Taxes will be redistributed to the unemployed in the form of a guaranteed minimum income (GMI).

- This will result in: stable economic growth.

A similar scenario will play out to protect against economic collapse due to a lack of resources like peak oil. Technological improvements and wide-scale automation will provide productivity improvements, not only because it lessens the use of human labor, but also because it drastically reduces the use of materials. Automation and technology also make it easier to mine, refine, and reclaim natural resources. The future will be a time of abundance, not scarcity. Technology is and always has been a deflationary counterbalance to the inflation.

Technology's Impact on Commodity Prices

Technological innovation can have drastic effects on commodity prices. If electric motors are successful in replacing automobile internal combustion engines, then lithium is a likely winner and platinum a loser. Lithium is used to manufacture batteries. Platinum is used in the catalytic converter that cleans gasoline engine exhaust.

Even if electric cars become the preferred mode of transportation, there is no guarantee that lithium prices will increase. Perhaps battery technology will derive more efficient use of lithium or replace it altogether with another material.

The recent collapse in oil prices is illustrative of how technology can drastically impact commodity prices. In 2008, the fear of peak oil production was very fashionable. The conventional narrative was that oil reserves were declining and fossil fuel energy demand was increasing. That year, West Texas Intermediate oil (which the US uses as an oil benchmark) hit a record high of $145.29 per barrel. Eight years later, on February 11, 2016, the price of oil had plummeted to $27. Throughout 2016, oil's average price would hover around $48. What happened to peak oil? Technology.

Energy use is becoming more efficient. LED lights, electric appliances, furnaces, automobiles, and everything else uses less energy. Additionally, alternative energy sources are becoming increasingly available. Renewable sources like solar, wind, and biomass now make up 7 percent of US electricity generation. Electric motors and battery technology are becoming affordable replacements for gasoline engines.

The mother of all energy-related technological innovation has been hydraulic fracturing (fracking) and directional drilling (also known as horizontal drilling). Fracking is a process where oil and natural gas are liberated from shale rock formations. High-pressure fluids are injected into subterranean shale, fracturing the rock. The fissures that are formed allow the previously locked-up oil and natural gas to freely flow, and thus be easily extracted. Oil reserves that were once considered economically unrecoverable are now being profitably produced for under $30 per barrel.

Traditional vertical drill bits consisted of "dumb" cutting teeth, which did little more than grind their way straight down into the earth. In directional drilling, the bit has been given intelligence through a complex robotic system that allows precise movement. The drill can be manipulated from the surface similar to the way a surgeon uses an endoscope during an operation.

Directional drilling and fracking are actually both old concepts. Attempts at directional drilling started in the 1920s, and fracking has been used since the late 1940s. The recent shale oil boom occurred because these old methods have been dramatically improved upon with advanced robotic systems that allow precision drilling of horizontal wells. The old drilling method was to bore one hole at a time over a prospective oil reservoir. If the well missed the oil patch, it was deemed "dry," and another hole had to be drilled.

Directional drilling techniques allow well holes to be drilled more than one mile deep and then extended out horizontally additional miles. The well head no longer needs to sit directly over an oil reserve. One main vertical well can be bored and then numerous horizontal shafts can extend outward to follow the contour of the reservoir.

Directional drilling is an economic game changer. Traditional drilling methods in deep ocean waters or other remote harsh environments are expensive and time consuming. A traditional deep water well can cost billions of dollars and take a decade to develop. Horizontal shale oil wells can be drilled at a lower cost and be operating in weeks.

Access to shale oil deposits has contributed to the US doubling production since 2008, making the US a top oil producer alongside Saudi Arabia and Russia. New oil finds and improvements in extraction techniques will continue to strengthen the US's capacity. In late 2016, a previously unknown oil reservoir was discovered in west Texas' Permian Basin. The shale oil formation known as Wolfcamp is estimated to contain 75 billion barrels of oil, which would make it the world's second largest next to Saudi Arabia's Ghawar Field.

The shale oil revolution has occurred because of the confluence of diverse technologies coming together and creating an entirely new approach, one that was not foreseen by the establishment. The companies that are dominating US shale oil production are not the exploration giants, like ExxonMobil or Chevron, but rather the smaller firms, like Pioneer, Devon, and Anadarko.

The technologies responsible for the economic recovery of shale oil are similar to the trends that have been discussed throughout this book. Advanced sensors and computer mapping software help locate geologic anomalies that might contain oil reservoirs. Massive big data cloud computing systems then process the exploration data, helping geologists pinpoint potential locations. The drillers at the wellhead use this data to steer the drill bit directly into the reservoir. Frackers can then inject the well with high-pressure fluids to liberate the oil from the shale.

All this technology has culminated into a competitive deflationary force that has cut the price of oil by more than 60 percent. I expect the trend to continue as fracking and directional drilling techniques are utilized outside of the US.

DRIVERLESS DUMP TRUCK

Rio Tinto is one of the world's largest commodity producers, operating mines on six continents. Recently, they have started using autonomous trucks to increase productivity at two iron ore mines in Australia. The trucks are completely automated and controlled remotely by employees at an operation center 700 miles away.

The trucks are not little Toyota pickups, but rather mammoth machines standing 29 feet tall and capable of hauling loads of 320 tons. There are currently 71 trucks in the fleet, operating 24 hours a day, seven days a week, and accounting for about 20 percent of the material flow.

Rio Tinto estimates they are saving at least 500 hours per year compared to using human drivers. Unlike humans, driverless trucks do not show up late, take vacation, require bathroom breaks, or get bored.

The projected savings far outweigh just replacing the drivers. Also eliminated is all the ancillary staff—supervisors, trainers, and all the employees needed to feed and care for the drivers at these remote mining locations. Add to that more consistency, lower incidence of accidents, less fuel consumption, and on and on. The savings are substantial enough that Rio Tinto plans to implement unmanned operations throughout their network of mines, including trains and drills.

The long-term goal in Australia is to have most of the supply chain operations from mining pit to shipping port run remotely. Bad news for the machine operators, but good news for the 400 humans that monitor operations at the control center in Perth, jobs that are not exposed to harsh environmental conditions and often dangerous accidents that occur at the remote mine sites.

The cost of other commodities is likely to remain suppressed as well. The technological advances that have made oil exploration more economical will provide similar benefits to mining other resources. Also, lower energy costs translate into overall lower commodity prices. That is because the highest cost of commodity mining and refining is energy. Diesel fuel is used to propel the large equipment that mines and transports the commodities. Coal or natural gas is used to power the refining process. For the same reasons, lower fossil fuel costs will have a similar impact on agricultural commodities. Lower energy costs are deflationary in nature and make everything less expensive.

The example of how technology has made oil exploration less expensive is not unique. Technology is always creating alternatives that favor one resource over another. In the nineteenth century, people lit their homes with candles made from tallow or whale oil. That was replaced by kerosene lamps, which were eventually replaced by incandescent lightbulbs, which are now being replaced with LED lights.

What does the future hold? Maybe peak oil. Or, perhaps, an abundance of energy from renewable sources, or maybe even clean nuclear fusion. New material might be invented that replaces scarce commodities or, perhaps, we will find a way to mine asteroids. No one can predict the future, but, clearly, energy prices in the present are leaning toward deflation.

From an investment standpoint, I personally would avoid large, long-term positions in commodities, particularly in any one commodity. If I were worried about inflation, I would hedge a small portion of my portfolio (5 to 10 percent) by purchasing a fund that invested in a basket of currencies.

Action Plan

1. I am interested in investing a portion (up to 10 percent) of my portfolio in commodities as a safeguard against inflation and to preserve purchasing power:

a. YES: Proceed to #2.

b. NO: This is probably a prudent decision, especially if my investment portfolio is not greater than $100,000.

2. As a starting point, I will study the following ETFs to see if they meet my criteria for investing in commodities (alphabetical listing by ticker symbol from various fund providers):

a. DBA (agriculture)

b. DBB (industrial base metals)

c. DBC (general commodities index)

d. GLD (gold)

e. SLV (silver)

f. USO (oil)

g. WOOD (timber and forestry)

Chapter 17

COMPANY OWNERSHIP (STOCKS AND ETFs)

In the future, as it has been in the past, the best source for building wealth will be ownership of a growing company. One hundred percent ownership in your own firm is an excellent course of action; thus, I have emphasized entrepreneurship throughout this book. However, as a matter of asset allocation, even those who have built extensive wealth in their own company should still have a portion of their net worth invested in common stocks. Also, common stock ownership may be the only option for some people because they do not want to be an entrepreneur or have not yet started their own business.

Company ownership provides the best source for building wealth because compensation is derived from more than just one's own individual effort. In a previous chapter, I described how an employee has less potential to build wealth because their only income stream is their paycheck. Conversely, their employer has four sources of compensation:

- His own effort

- The effort of the employee

- Rent on his land

- Return on the use of his capital

Ownership in company stock provides similar rewards but without having any managerial or employee responsibility. Stock ownership makes an investor a silent partner that is able to share in the rewards of a company's growing profits. The operative words there are *growing* and *profits*. Nothing else really matters.

Of all the asset classes, I believe stock ownership will provide the best future investment opportunities. While real estate and commodities are likely to keep up with inflation, companies that position themselves to take advantage of automation will make hoards of money. Robots increase productivity, productivity increases profits. It is that simple.

A stock's price is ultimately derived from its earnings. The reason there is so much fluctuation in price is that despite conventional wisdom, the market is not efficient. Future earnings are always uncertain; they could rise or fall because of a change in operating methods, fickle consumer preferences, weather conditions, or just plain luck.

Estimates of future earnings drive investors to place a premium or a discount on the stock's valuation. If profits are expected to grow, investors will bid the stock price up; if profits are expected to decline, investors will bid the stock price down. Essentially, stock prices swing above or below the actual value of the company based on investor fear and greed.

The common denominator of stock price appreciation is the expectation of increasing future profits. This has always been the case. It will be the force that drives stock prices in the future. In fact, it will be even more critical in the robotic future because of the speed that new technologies will be adopted and the likely failures of established companies once they become victims of competitive disruptive innovation. The stock market is likely to be highly volatile and offer both the best and worst of opportunities.

PROFITS WITHOUT PEOPLE

Just as the digital age has made information virtually free, automation will all but make labor free. Human labor represents approximately 60 percent of a corporation's cost of doing business. As people are replaced by machines, corporate profits will rise while product prices decline.

This phenomenon has been occurring for at least a decade. For the past fifteen years, the corporate profits of the S&P500 have risen approximately 10 percent per year, while at the same time, sales revenue has barely grown 1 percent annually. While some of this profit expansion can be attributed to balance sheet shenanigans, most of it is due to increases in productivity, a euphemism for replacing human labor with automation. Cynics have dubbed this transition as "profits without people."

So how can a stock investor not only navigate the turbulent times ahead, but profit from them? Below I will present some concepts and strategies to help sharpen your investing wit. Several of these will be counterintuitive or go against the grain of conventional wisdom. It has been my experience that taking a contrarian view is often a good strategy to prevent becoming too overconfident, particularly when chasing a popular trend.

I will be referring a lot to the stock market. I use this as a catchall phrase to mean trading stocks and exchange-traded funds (ETFs). Since this can represent individual company ownership, indexes, specific sectors, bonds, foreign issues, commodities, and real estate, the stock market literally provides a means to own virtually anything. I am a huge proponent of investing via the stock market because it provides a simple means of asset diversification to even the smallest of portfolios. There has literally never been a better time in human history to be an investor.

MYSPACE

In 2005, media mogul Rupert Murdoch was convinced social media would take off. The multibillionaire purchased Myspace for $580 million. Many of you have never heard of Myspace because it was an abysmal failure.

In spite of the fact that Murdoch is a business and media genius and Is no doubt surrounded by the brightest of financial minds, he bet on the wrong horse.

Picking winners is not easy. Learn to discount your survival bias and make sure to diversify your holdings so that all your eggs are not in one basket. Unless you are smarter than Rupert Murdoch, you are sure to pick some losers.

Stock or ETF?

There are numerous ways to invest in equities (common stock): There's direct ownership of company stock, like purchasing shares of Apple (AAPL), or indirect ownership through funds such as mutual funds or ETFs.

Mutual funds were revolutionary in their day, but in my opinion, they will soon be relegated to the dustbin of history. Their obsolescence is being driven by ETFs that are less expensive and offer more choice.

Mutual funds are run by investment companies that were structured under the antiquated financial laws of 1940. Whenever an individual investor purchases or redeems shares of a mutual fund, the fund is required to rebalance its portfolio to reflect the change in net asset value. Rebalancing can result in a great deal of inefficient trading, which increases the fund's transaction costs, record keeping, taxable income, and regulatory compliance. All these costs must be passed onto individual investors in the form of fees or reduced gains.

In spite of lower cost alternatives like ETFs, the historic stature of mutual funds has kept them deeply entrenched within established financial service companies, hence their near monopolistic use in corporate 401(k) plans. I only recommend mutual fund ownership to investors that are forced to use them due to limitations of restrictive retirement programs, like most employer sponsored 401(k) plans.

ETFs are similar in nature to mutual funds, allowing small individual investors to indirectly own shares of companies, trusts, or equity derivatives. The underlying difference is in the more efficient mechanism ETFs use to purchase assets for individual investors.

Rather than pooling investor money and having the fund be responsible for making individual asset purchases, ETFs use a sophisticated system of third party "authorized participants." Each authorized participant competes against the others to fill orders and thus balance the fund's overall net asset value. This competitive trading environment encourages each participant to maximize their profit potential by executing trades in the most cost-effective manner. How do they accomplish this task? By using "trading robots" that are powered by high-speed computer algorithms.

The ETF benefits from not only the lower overall cost of trading, but also from passing the responsibility and associated paperwork onto the authorized participants. The ETF firm can operate very cost effectively and thus charge lower fees to the individual investor. For example, an ETF's management fee is generally at least 20 basis points (0.2 percent) cheaper than a comparable mutual fund, which represents about a 70 percent savings for the consumer.

In addition to lower costs, the trading efficiency of ETFs has also resulted in an extensive offering of investment sectors. ETFs that mimic large indexes like the S&P500, or niche sectors like coffee bean futures, can be purchased. Even if ETFs did not represent a cost advantage over mutual funds, I would use them as a means to effectively diversify my investment portfolio.

ETF MINUTIA

The mutual fund industry has a history, dating back over 75 years. Their incumbent position as the financial industry's preferred choice for corporate 401(k) retirement plans is virtually unchallenged. In the US, there are over 9,000 mutual funds managing in excess of $15.5 trillion. But their prestige is diminishing. Investor money is flowing out of mutual funds and into ETFs at an escalating rate. Recent trends show mutual funds losing $17 billion per month.

The fight between ETFs and the established mutual fund industry is a classic tale of David versus Goliath. All the large mutual fund investment companies now offer their own branded version of ETFs. As the ETF mode gains more market share, I expect the corporate structures will morph into an indistinguishable hybrid version of the two types of funds.

What will the end result be? Fewer human jobs in the financial industry and lower costs for investment products.

Whether you choose to invest in individual stocks or ETFs is not important. Both are listed on the major stock markets and can easily be purchased through a discount broker. Generally, I think ETFs should play a larger role for people who have less money to invest. For example, someone with a $5,000 portfolio might put the entire amount in one ETF that tracks the S&P500. Someone with a $100,000 portfolio might split the money equally into four ETFs that track large cap stocks, small cap stocks, international stocks, and bonds. A millionaire might try to increase profits by owing the same four ETF sectors and then supplementing with smaller positions in specific growth stocks.

The good news about investing in individual stocks and ETFs is that they are not limited to only owning equity assets. An investor can use stocks and ETFs to diversify into the other asset classes previously discussed: bonds, real estate, and commodities. For example, a share in precious metals could be achieved by purchasing the stock of a gold mining company, purchasing a variety of ETFs that own mining stocks, or the actual metal itself. Investing in real estate could be done in a similar manner through an ETF real estate investment trust (RIET). Bonds can be owned through an ETF bond fund.

Personally, I would rather own a stock or ETF than invest directly in the underlying asset, like gold or real estate. If I own physical gold, I have to worry about storing and safeguarding it. If I own real estate, I become a landlord. By owing an ETF, I can get investment exposure to any of these asset classes with no hassle or fuss.

Mitigating Specific Stock Risk

Using sector ETFs is also an extremely effective method for mitigating risk, especially when it comes to investing in the more speculative areas of the market, like technology. Rather than putting all your cybersecurity asset allocation into one stock, like Palo Alto Networks, CyberArk, FireEye, or Proofpoint, why not take a small bet on all of them by purchasing an ETF that specifically invests in the entire cybersecurity space? Using sector-specific ETFs is one of my favorite methods for taking advantage of a trend.

HACK is the stock symbol for an ETF that invests solely in the cybersecurity sector. Purchasing this fund would provide exposure to all the leading global cybersecurity companies. So rather than limiting your choices by investing in one or two superstars, like Palo Alto or FireEye, owning HACK would represent a small share in 100 different cybersecurity companies.

As mentioned previously, ETFs provide a very cost-effective means to easily diversify a portfolio. In the case of HACK, by purchasing this one fund, in addition to broad exposure to 100 cybersecurity stocks, the fund also provides concentrated positions in the sector leaders. The fund is managed, so as new company leadership emerges, the fund rebalances its position in the winning stocks.

HACK's current top ten stock holdings represent 46 percent of the fund's overall investment. Portfolio allocation between these ten funds is split approximately equally, at 4.6 percent per company. Palo Alto Networks, Proofpoint, and Symantec are represented among these holdings. So, a prudent investor seeking ownership in a growing technology sector could easily diversify their portfolio by investing 10 percent of portfolio value in HACK. This one purchase would provide broad exposure to 100 cyber-security companies with a slightly concentrated position in the sector's leadership. Thus, a 10 percent portfolio allocation to HACK would provide a very conservative 0.46 percent specific stock risk to superstar performers like Palo Alto Networks.

Using ETFs to diversify into specific industry sectors does not eliminate the risk of catastrophic loss; it only mitigates the risk. During the 2000 dot-com bubble and the 2008 financial crisis, the NASDAQ 100 composite ETF named QQQ lost 82 percent and 50, percent respectively. In nominal terms (non-inflation adjusted), it took QQQ 15 years to recover from the dot-com loss. ETFs can and do lose money, sometimes significant amounts.

There is a sector-specific ETF for almost every conceivable market sector. New ETFs are constantly coming onto the market. In 2013, an ETF was launched that invests specifically in robotics and automation companies. Its stock symbol is appropriately named ROBO.

As might be imagined, ROBO is a high-risk investment. Investors with a long-term horizon will likely be rewarded but, in spite of the catchy name, its performance has not been stellar. Since inception, its overall gain has been a little more than 10 percent, a time when a much "safer" investment in the S&P500 would have produced a 25 percent gain. High-risk investments are not guaranteed to produce above-average results. More times than not, you can count on them to produce below-average results, which is why I stress prudence and portfolio diversification.

MY FAVORITE ROBOT STOCK

If I had to choose just one individual automation company to invest in for the long term, it would be Switzerland-based ABB. Its autonomous motors, drives, and controllers are among the best quality in the world. Additionally, the company has a consistent history of making profits since its founding in 1883. ABB is essentially Switzerland's equivalent of General Electric (GE). It has deep roots in the best sectors of industrial products, power generation, and transportation. Over the past 133 years, ABB has survived world wars, depressions, and all manner of technological innovation. I suspect it will survive and thrive in the robotic future.

However, that does not mean that ABB is risk free or that its performance will always be stable. Although I believe ABB, like America's GE, is a good bet for reaping profits in the robotic future, technologically advanced companies are not necessarily more profitable than their low-tech counterparts. Case in point, since 2007 both ABB and GE have significantly underperformed the S&P500 broader market. In fact, over that same time period, technologically advanced ABB has underperformed the boring Packaging Corporation of America (stock symbol PKG) by well over 200 percent.

So how can a manufacturer of low-tech products like paper boxes outperform an industrial powerhouse like ABB, and even the broader general market? It all comes down to profits.

A paper box has not changed form much over the past 100 years, but the manufacturing process that creates it has. A non-flashy company like PKG may not develop innovative new technology, but it consumes and implements new manufacturing techniques. Since technology allows old things to be made more efficiency, PKG can keep squeezing more profits out of old product lines, like paper boxes.

Since the relative price of packaging remains moderate, those old boring boxes keep finding new uses, like shipping all that stuff you order from Amazon to your front door.

Ancillary Winners

Another way to mitigate risk is to invest in companies that implement technology rather than those that create it. I have no idea which company will manufacture the best robot a decade from now. The company may not even exist today. What I can be pretty sure of is that regardless of who produces the best robot, companies like Procter & Gamble (P&G) will use that technology to improve their profitability. Lower labor costs and more efficient manufacturing techniques will mean that the consumer staples that P&G manufactures will all get cheaper. Some of that cost savings will be passed along to consumers, a portion of it will bolster P&G's bottom line, and some of it will go to the product development of better toothpastes and soaps. In any case, P&G and their shareholders will benefit.

Mitigating risk by focusing on ancillary technology winners can also be achieved by investing in companies that are upstream in the supply chain from the riskier innovators. Driverless autonomous cars will likely be a consumer product of the future. Does that mean that Tesla will be a better long-term investment than Hyundai? Probably not, because a low-cost manufacturer like Hyundai will adopt autonomous driving technology and affordably mass produce it in volume just like all their other car features.

Looking upstream beyond the car manufacturer will provide some interesting investment opportunities. Start with a Tier I preferred automobile parts manufacturer, like Delphi, that supplies to many car companies. Delphi makes an assortment of driving technologies from powertrains to electronic components. Recently, Delphi has teamed up with Mobileye to launch an autonomous driving system that can be mass produced to fit a variety of car manufacturer's platforms, not just fancy, expensive Teslas. Delphi is a profitable company and might serve as a conservative approach for an investor looking for at least some exposure to the autonomous car trend.

For those that have a high-risk tolerance, perhaps look at Mobileye. Mobileye is an Israeli-based company that makes the camera-based sensor that was initially used by Tesla. If Mobileye's partnership with Delphi is successful, the system would have mass appeal and provide much higher revenue than an exclusive deal with Tesla. Mobileye is currently profitable but has a very high valuation and a great deal of risk specific to autonomous driving systems, because that is its main offering. Perhaps that is too risky for you.

Look further into the autonomous driving supply chain and you will find Nvidia. They have also recently ventured into the autonomous car space with a new device that claims to be a mini-supercomputer capable of rapid self-learning. This is exactly the type of artificial intelligence (AI) needed by autonomous cars so that they can quickly react to changing conditions. Nvidia is probably less risky than a pure play like Mobileye, because even if Nvidia's new device does not pan out for autonomous vehicles, it likely has other AI applications. Furthermore, Nvidia is profitable and has a diverse business designing graphic processing units for applications like computer workstations and gaming consoles.

Perhaps Nvidia is still too risky for you and you would prefer exposure to the autonomous vehicle space with a larger, more stable technology company. Then consider Intel, one of Silicon Valley's original semiconductor manufacturers. Intel is an extremely large company with a market capitalization of $160 billion, so they would not have the same growth potential as an Nvidia or a Mobileye. However, for that very same reason, Intel is stable and likely to profit from all the future automation trends. Intel also has an extremely moderate valuation compared to other technology companies.

I took you through the above exercise not as a specific recommendation for what stocks to buy in the autonomous driver space, but rather as an illustration of the process. Having the ability to look up and down the supply chain for ancillary opportunities is a valuable skill set. The same process will work for any sector of the stock market, not just technology. It will help you sift through the noise of media-hyped stocks and avoid "hot tips." I particularly like using this method because it can be adapted for any level of desired risk.

Final Word on Stocks

I want to close out this chapter by reiterating the importance of owner-ship in a company, whether through a personal small business or the stock market. Throughout history, the best method of wealth creation has been achieved through ownership of a growing enterprise. That is a trend that will continue far into the robotic future.

If you have the inclination for entrepreneurship, then by all means build your own business. Entrepreneurship provides a dual path to both wealth and happiness. The next best alternative that I know of is ownership via common stocks. The stock market provides the means to share in the suc-cess of the best companies in the world, with comparatively little risk and effort. It is not foolproof, but it is attainable.

There has never been a better time to be an individual investor. Humanity is about to enter a period of epic change brought on by the creative dis-ruption of automation. Corporations that are able to successfully navigate through these uncharted waters will be rewarded with an unprecedented concentration of wealth. The bounty will be shared with those intrepid individuals who are willing to invest. I encourage you to at least consider the possibility.

Action Plan

1. I have at least $50,000 to invest (from all sources: savings, 401k, IRA, etc.):

   a. YES: Proceed to #2.

   b. NO: I probably do not have enough capital to make investing worth the effort and should instead focus my efforts on earn-ing and saving more money.

   Note: This is an extremely personal decision, but consider the person who only has $10,000 to invest. If they were to achieve a 10 percent return, that would be a profit of $1,000. Their time and effort might have been better spent in other income-producing activities, like increasing their skills, working overtime, or seeking a higher paying job.

2. Before investing my hard-earned money, I will re-read this chapter and pay particular attention to the story about Rupert Murdoch's poor investment in Myspace. I will begin to invest once I fully understand the consequences of my decisions.

THE ROBOTS ARE COMING

The robots are coming; are you prepared? For those who are not, unhappiness likely awaits. We cannot be sure when the harsh impact will be felt, or if the pain will come abruptly or like a slow burn. What we do know is that unprecedented change is on the way. Indeed, the early warning signs have been flashing for several decades. Global growth has stagnated and will likely decline further. As labor becomes near free, the common employee's future will be dim.

The robots are coming to take the jobs of human workers that perform routine tasks. Not just the assembly line worker or the fast food restaurant employee, but especially the smug white collar professional; all those whose primary function can be reduced to a mathematical algorithm. The highly compensated professionals will not be spared, but instead, their confiscatory incomes will provide the market with a catalyst to replace them with automation.

The robots are coming to enable people, especially the disadvantaged. While many will be made redundant or forced into early retirement, others will become empowered. The difference between the winners and losers will be application versus fear. The winners will embrace automation and use its enabling technology to overcome their weaknesses and improve their strengths. The losers will fear robots and shun technology.

Both winners and losers know the same thing about automation: People cannot outperform robots at routine tasks. Robots are faster, cheaper, and more productive than humans. The physical stamina of robots has outperformed us for some time. Now, robotic intellect is sharpening. Automation aided by inexpensive miniature supercomputers will have near-infinite memory. Artificial intelligence of the average robot will soon outstrip the cognitive abilities of the average human.

Optimism lies in the four principles of thought that are described in this book. They will help guide you through the turbulent economic times that will be brought on by ubiquitous automation. The power to reign supreme over the robot is internal to each of us. The battle will not be won by physical competition but by how we think.

Think like a human, not a machine. Your economic worth is derived from the ability to create, not to perform repetitive tasks. Develop the human touch skills that are unique to your personal talents and abilities. Your emphasis should be on creation.

Think like an entrepreneur, not an employee. Unemployment will most likely reach epidemic levels as people are replaced by more efficient machines. Jobs may be few, but opportunity will abound for those that have learned to monetize their human touch.

Think like a saver, not a consumer. The emphasis of automation is on more efficient production, creating a deflationary environment where prices fall. Deflationary forces favor savers that have the discipline to postpone gratification. Consumers buy to satisfy emotion, savers identify value and purchase appreciating assets.

Think like an investor, not a speculator. The creative destruction of technology will obsolete many established industries, creating amazing investment opportunities. Filtering through the noise of media-hyped bubbles will be a critical skill to safeguard capital. The prudent investor will find ample ways to magnify his wealth.

So how do you stack up? Are you prepared to take on the robots? If not, do not despair. You are ahead of the game and there is still time to prepare.

Go back and reread the chapters of the book where your skills are lacking. Worry less about what you cannot do and focus on your innate human touch abilities. Technology will be available to circumvent the things you are not good at. Focus your efforts on improving your unique talents that enable you to create.

Once you have looked inward to develop your strengths, then look outward toward others and to the economy. Let the market tell you what is needed. Review the sections in the book where you filled out action plans. This is an iterative process, so do not hesitate to make changes and reconsiderations. Your course will naturally change as new information comes to light or as your self-discovery unveils new talents.

Start with where you are and what you have. Identify your unique human touch and your personal character traits. Pursue one or more leading traits in your general area of interest. As you become more proficient in that field, your confidence will grow and you will be able to develop more of the supporting traits. Your life will be in balance and you will be more likely to provide the marketplace with creative products and services that are in high demand.

There is no one path or direct route to success. The future is always uncertain, and the technological advances of automation are disrupting things at an unprecedented rate. The future has never been murkier.

As you make your journey into the uncharted robotic economy, recall the Lewis and Clark Expedition of 1804. They had no definitive map to follow, only a general direction to pursue. They prepared by developing skills that would be useful for the journey. As they made their way and encountered changing conditions, they improvised and adapted. You will do well to follow their example.

Our modern-day advantage over the Corps of Discovery is that we do not have to make the expedition alone. As you proceed on your journey, I encourage you to follow along with me at my podcast's website: www.wealthsteading.com

The podcast is a forum where I discuss current market conditions and provide insight into general wealth-building principles, topics intertwined with the inevitable robotic economy of the future. So, please join me for updates as well as let me know how you are progressing.

In spite of the gloom and doom predicted by many prognosticators, I remain optimistic about the future, especially for those of us who are prepared. We are fortunate to live in amazing times. Being proactive and changing the way you think will be key to your future prosperity.

As you encounter difficulty, do not despair. Your efforts will be rewarded and, hopefully, like Lewis and Clark, we will all get lucky and recognize our Sacagawea. So be of good cheer...the robots are coming to make you wealthy!

INDEX

ABOUT THE AUTHOR

John Pugliano approaches the subject of automation from two distinct vantage points. He is the founder and money manager of Investable Wealth LLC, an independent investment advisory firm. In that role, John uses his thirty years of investor experience to identify trends and translate their significance into profitable stock market trades. He has long followed the creative, destructive forces of automation and how they impact the economy.

Additionally, in a former career, John spent 20 years involved in the sales, marketing, and product development of industrial products. He started his career as an industrial sales representative in the Midwest, where he witnessed firsthand the competitive forces of automation and globalization that led to the closure of many US factories. As his career progressed, he followed the displacement of manufacturing from the US to production facilities across the globe.

A student of history and a keen observer of human nature, John uses examples of technology's impact on the past to weave an interesting story of steps we can take to success in the robotic future. The result is a blend of cynical realism tempered with hopeful optimism for the future of humanity.